Selected Titles in This Series

160 **V. G. Osmolovskiĭ,** Linear and nonlinear perturbations of the operator div, 1997

159 **S. Ya. Khavinson,** Best approximation by linear superpositions (approximate nomography), 1997

158 **Hideki Omori,** Infinite-dimensional Lie groups, 1997

157 **V. B. Kolmanovskiĭ and L. E. Shaĭkhet,** Control of systems with aftereffect, 1996

156 **V. N. Shevchenko,** Qualitative topics in integer linear programming, 1997

155 **Yu. Safarov and D. Vassiliev,** The asymptotic distribution of eigenvalues of partial differential operators, 1997

154 **V. V. Prasolov and A. B. Sossinsky,** Knots, links, braids and 3-manifolds. An introduction to the new invariants in low-dimensional topology, 1997

153 **S. Kh. Aranson, G. R. Belitsky, and E. V. Zhuzhoma,** Introduction to the qualitative theory of dynamical systems on surfaces, 1996

152 **R. S. Ismagilov,** Representations of infinite-dimensional groups, 1996

151 **S. Yu. Slavyanov,** Asymptotic solutions of the one-dimensional Schrödinger equation, 1996

150 **B. Ya. Levin,** Lectures on entire functions, 1996

149 **Takashi Sakai,** Riemannian geometry, 1996

148 **Vladimir I. Piterbarg,** Asymptotic methods in the theory of Gaussian processes and fields, 1996

147 **S. G. Gindikin and L. R. Volevich,** Mixed problem for partial differential equations with quasihomogeneous principal part, 1996

146 **L. Ya. Adrianova,** Introduction to linear systems of differential equations, 1995

145 **A. N. Andrianov and V. G. Zhuravlev,** Modular forms and Hecke operators, 1995

144 **O. V. Troshkin,** Nontraditional methods in mathematical hydrodynamics, 1995

143 **V. A. Malyshev and R. A. Minlos,** Linear infinite-particle operators, 1995

142 **N. V. Krylov,** Introduction to the theory of diffusion processes, 1995

141 **A. A. Davydov,** Qualitative theory of control systems, 1994

140 **Aizik I. Volpert, Vitaly A. Volpert, and Vladimir A. Volpert,** Traveling wave solutions of parabolic systems, 1994

139 **I. V. Skrypnik,** Methods for analysis of nonlinear elliptic boundary value problems, 1994

138 **Yu. P. Razmyslov,** Identities of algebras and their representations, 1994

137 **F. I. Karpelevich and A. Ya. Kreinin,** Heavy traffic limits for multiphase queues, 1994

136 **Masayoshi Miyanishi,** Algebraic geometry, 1994

135 **Masaru Takeuchi,** Modern spherical functions, 1994

134 **V. V. Prasolov,** Problems and theorems in linear algebra, 1994

133 **P. I. Naumkin and I. A. Shishmarev,** Nonlinear nonlocal equations in the theory of waves, 1994

132 **Hajime Urakawa,** Calculus of variations and harmonic maps, 1993

131 **V. V. Sharko,** Functions on manifolds: Algebraic and topological aspects, 1993

130 **V. V. Vershinin,** Cobordisms and spectral sequences, 1993

129 **Mitsuo Morimoto,** An introduction to Sato's hyperfunctions, 1993

128 **V. P. Orevkov,** Complexity of proofs and their transformations in axiomatic theories, 1993

127 **F. L. Zak,** Tangents and secants of algebraic varieties, 1993

126 **M. L. Agranovskiĭ,** Invariant function spaces on homogeneous manifolds of Lie groups and applications, 1993

125 **Masayoshi Nagata,** Theory of commutative fields, 1993

124 **Masahisa Adachi,** Embeddings and immersions, 1993

123 **M. A. Akivis and B. A. Rosenfeld,** Élie Cartan (1869–1951), 1993

122 **Zhang Guan-Hou,** Theory of entire and meromorphic functions: deficient and asymptotic values and singular directions, 1993

(*Continued in the back of this publication*)

Linear and Nonlinear Perturbations of the Operator div

Translations of
MATHEMATICAL
MONOGRAPHS

Volume 160

Linear and Nonlinear Perturbations of the Operator div

V. G. Osmolovskiĭ

American Mathematical Society
Providence, Rhode Island

В. Г. Осмоловский

ЛИНЕЙНЫЕ И НЕЛИНЕЙНЫЕ ВОЗМУЩЕНИЯ ОПЕРАТОРА div

ИЗДАТ. С.-ПЕТЕРБУРГСКОГО УНИВ.
САНКТ-ПЕТЕРБУРГ, 1995

Translated from the Russian by Tamara Rozhkovskaya
with the participation of Scientific Books (RIMIBE NSU),
Novosibirsk, Russia

1991 *Mathematics Subject Classification.* Primary 35F15, 35F30; Secondary 35Q35.

ABSTRACT. This book presents the theory of boundary value problems for the operator div and its linear and nonlinear perturbations. Applications to geometry, the calculus of variations, and continuum mechanics are described. The book can be used by research mathematicians and graduate students working in partial differential equations and applications to mathematical physics.

Library of Congress Cataloging-in-Publication Data

Osmolovskiĭ, V. G. (Viktor Grigor′evich), 1947–

[Lineĭnye i nelineĭnye vozmushcheniia operatora div. English]

Linear and nonlinear perturbations of the operator div / V. G. Osmolovskiĭ; [translated from the Russian by Tamara Rozhkovskaya].

p. cm. — (Translations of mathematical monographs ; v. 160)

Includes bibliographical references.

ISBN 0-8218-0586-X

1. Boundary value problems. 2. Perturbation (Mathematics) 3. Mathematical physics. I. Title. II. Series.

QA379.08613 1997

515′.724—dc21 96-40489 CIP

---

10 9 8 7 6 5 4 3 2 1 02 01 00 99 98 97

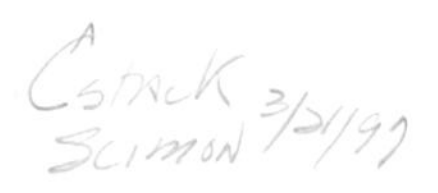

Contents

Preface

In various questions of analysis, it is often necessary to describe the set of those solutions of a nonlinear problem that lie in a sufficiently small neighborhood of a known solution. The present book is devoted to problems of this type in the case of a scalar first order differential equation for a vector-valued function. As a typical example, we mention the following problem:

$$\det \dot{y}(x) = 1, \quad x \in \omega \subset \mathbb{R}^m,$$
$$y(x) - x = 0, \quad x \in \Gamma \subset \partial\omega,$$

where $y(x)$ is an m-dimensional vector-valued function and $\dot{y}(x)$ is the matrix of the first derivatives of y. It is required to describe the set of those solutions of this problem that lie in a sufficiently small neighborhood of the known solution $y(x) \equiv x$.

A natural tool for studying such problems is perturbation theory. Applying it, we necessarily linearize the nonlinear equation near the known solution. In the above example, the operator div is a linearization and the operator det can be regarded as a small nonlinear perturbation of the operator div.

For the general nonlinear equation $F(\dot{y}, y, x) = 0$, a linearization is $Lu = C_{ij}u^i_{x_j} + C_i u^i$, a first order operator. Under some restrictions on the coefficients $C_{ij}(x)$ and $C_i(x)$, the operator L can be considered as a compact perturbation of the operator div.

The book discusses three topics.

1. Exposition of the theory of boundary-value problems for the operator L.

2. Exposition of the theory of nonlinear perturbations of the operator L, which allows us to give a local description of the set of all solutions to the boundary-value problem for the equation $F(\dot{y}, y, x) = 0$ for which L can be taken as a linearization.

3. Applications of the results.

Among scalar first order differential operators acting on vector-valued functions, of particular interest is the operator div which, among other properties, expresses the incompressibility condition in hydromechanics. We present linear and nonlinear perturbation theory for the operator div. The results are illustrated by various applications.

The book consists of two chapters and an appendix. In the first chapter, we consider the operator L that acts on n-dimensional vector-valued functions $u(x)$, $x \in \omega \subset \mathbb{R}^m$, by the formula

$$Lu(x) = \operatorname{div}\big(A^*(x)u(x)\big) + \big(a(x), u(x)\big)_n,$$

where $A(x)$ is an $(n \times m)$-matrix-valued function, $a(x)$ is an n-dimensional vector-valued function, ω is a bounded domain, and $(\cdot,\cdot)_n$ denotes the inner product in $\mathbb{R}^n$.

The entries of the matrix $A(x)$, components of the vector $a(x)$, and the boundary of ω are assumed to be sufficiently smooth; moreover, $\det A^*(x)A(x) \neq 0$ for $x \in \overline{\omega}$. In the case $n = m$, $A(x) \equiv I$, $a(x) \equiv 0$, the operator L becomes the operator div. The literature on the operator div is quite extensive. We mention only [**2, 3, 14, 16, 17**]. To study boundary-value problems for the operator div, we use a traditional method based on the identity div rot $= 0$. The general operator L is considered as a compact perturbation of the operator div. This allows us to avoid the application of Fredholm complexes reducing the problem to a form which admits the immediate use of the Fredholm alternatives. For the operator L we consider various boundary-valued problems and find necessary and sufficient solvability conditions in explicit form. In particular, the problem

$$Lu = f, \qquad u\big|_{\partial\omega} = \varphi$$

is solvable for all pairs of functions f, φ of a certain smoothness, while for the operator $L =$ div the well-known consistency condition serves as a solvability criterion. Particular attention has been given to the kernel $N = \{u : Lu = 0\}$ of the operator L. We prove an analog of the Weyl decomposition and show its stability. We emphasize that N is a natural generalization of the space of solenoidal vector fields. To study the operator L, we essentially use the description of the set of solutions to the problem

$$L^*p \equiv -A\nabla p + ap = 0, \qquad p \in C^1(\overline{\omega}).$$

Due to the very simple form of this problem, the description can be obtained without using the technique of overdetermined systems.

Using the results of the first chapter, we can classify the sets of all solutions to various boundary-value problems for the scalar nonlinear equation

$$F[y] = F(\dot{y}, y, x) = 0, \qquad x \in \omega,$$

where a mapping $y(x)$ of $\omega \subset \mathbb{R}^m$ into $\mathbb{R}^n$ with the Jacobi matrix $\dot{y}(x)$ belongs to a sufficiently small neighborhood of the known solution $z(x)$; this can be done provided that the linearization of the operator $F[y]$ on $z(x)$ is given by the operator L whose coefficients satisfy the conditions formulated above. The classification and applications in geometry, the calculus of variations, and incompressible continua are discussed in Chapter 2.

To describe the set of all those solutions to the problem $F[y] = 0$ that lie in a small neighborhood of the known solution $y = z$, it is natural to use the implicit function theorem or the implicit function theorem together with the Lyapunov–Schmidt splitting procedure. We can manage with the implicit function theorem alone if the problem $L^*p = 0$ has only the zero solution; otherwise, it is necessary to use, in addition, the Lyapunov–Schmidt splitting procedure. If $L^*p = 0$ has only the zero solution, then the set of solutions has the structure of a surface in a small neighborhood of the point z in some function space; moreover, N is the tangent space to this surface at the point z. Otherwise, there are variants depending on properties of the bifurcation equation and the boundary conditions imposed on $y(x)$. In particular, two extreme cases can occur. In the first case, the set of solutions looks like a piece of surface in some function space and N is the tangent space to this surface at the point z. In the second case, $y(x) \equiv z(x)$ is an isolated solution. The boundary conditions or some other additional conditions imposed on $y(x)$ are called rigid if the set of solutions to the problem $F[y] = 0$ consists of an isolated solution $y(x) = z(x)$. This term is introduced by analogy with the terminology in the problem about isometric deformations (cf. the historical survey in [**1**]). Particular emphasis is placed on studying the rigidity of various conditions.

As examples of $F[y]$ we consider operators such that $F(\dot{y}, y, x)$ coincides with an invariant of the metric tensor $g_y = \dot{y}^* \dot{y}$ of the mapping $y(x)$. The most popular such problem is that of describing the set of all those solutions to the equation $\det g_y - 1 = 0$, $x \in \omega$, $y(x) - x = 0$, $x \in \partial\omega$, that lie in a sufficiently small neighborhood of the solution $z(x) \equiv x$. Such a problem often occurs in continuum mechanics.

Themain part of the second chapter is devoted to variational problems for the functional with constraint

$$J[y] = \int_{\omega} H(\dot{y}, y, x)\, dx + \int_{\partial\omega} h(y, x)\, dS, \qquad F(\dot{y}, y, x) = 0.$$

For the existence conditions for a global minimum of this functional we refer to [**4, 22**]. We will touch only on necessary conditions for an extremum. A general approach to the search for necessary conditions for an extremum in problems with constraints is discussed, e.g., in [**13, 20**]. Information about the structure of the set of solutions to the problem $F[y] = 0$ in a neighborhood of the extremal point z allows us, following the general approach, to derive the Euler–Lagrange equation, compute the second variation of the functional, obtain the Legendre–Hadamard conditions, and justify the method of Lagrange multipliers. We note that under the incompressibility condition, Lagrange multipliers were applied for solving variational problems in [**10, 11**]. To obtain sufficient conditions for a local minimum over sufficiently small smooth perturbations, we use the multiplicative inequalities (cf. the Appendix).

For statements of variational problems in continuum mechanics we refer to [**19**]. In particular, if $J[y]$ is the energy functional of deformations of an elastic body, then for $F[y] = \det \dot{y}(x) - 1$ the problem of minimizing the functional $J[y]$ with the constraint $F[y] = 0$ is a problem about the equilibrium state of an elastic incompressible medium.

As for prerequisites, the reader is expected to be familiar with basic notions and facts from functional analysis and the theory of elliptic differential equations (cf., e.g., [**5, 7, 8, 12, 18, 20, 21, 23, 35–37**]). For the convenience of the reader, in Chapter 2 we present the necessary facts from the differential calculus in normed spaces and from bifurcation theory, following [**5, 12, 23, 35**]. Necessary information about function spaces, extension theorems, and multiplicative inequalities is contained in the Appendix.

The main results presented in the book are based on the author's papers [**24–34**]. We note that the choice of the material included in the book was conditioned by the author's interests. It is not our purpose to present all current results on boundary-value problems for scalar first order equations. In particular, we do not discuss such questions as nonstationary problems, quasilinear first order equations, equations with nonsmooth coefficients, and domains with nonsmooth boundaries.

Notation

Euclidean space

The summation convention over repeated indices is sometimes assumed.

$\mathbb{R}^m$ is m-dimensional Euclidean space, with inner product $(\cdot,\cdot)_m$ and modulus (absolute value) $|\cdot|$.

$\mathbb{R}^m_+ = \{x = (x_1, \dots, x_m) \in \mathbb{R}^m : x_m \geqslant 0\}$.

$B_\rho(x_0) = \{x \in \mathbb{R}^m : |x - x_0| < \rho\}$.

Domains in Euclidean space

ω is a bounded domain in $\mathbb{R}^m$ with closure $\overline{\omega}$ and boundary $\partial\omega$.

Γ is an open subset of $\partial\omega$ with closure $\overline{\Gamma}$.

$S = \partial\omega/\overline{\Gamma}$.

$\rho(x)$ is the distance from x to $\partial\omega$.

$d(x)$ is the oriented distance from x to $\partial\omega$ (cf. the Appendix).

$\nu(x)$ is the unit outward normal to $\partial\omega$.

h_ν is the normal component of a vector-valued function h defined on $\partial\omega$.

Function spaces

$L_q(\omega)$, $q \in [1,\infty)$, is the space of q-integrable functions in ω, equipped with norm $\|\cdot\|_q$.

$W^l_q(\omega)$, $l \geqslant 0$, $q \in [1,\infty)$ (the Sobolev–Slobodetskiĭ space), is the space of functions that, together with their Sobolev derivatives of order up to l (fractional l is possible), are q-integrable in w, equipped with the norm $\|\cdot\|_{l,q}$.

$C(\overline{\omega})$ is the space of continuous function defined in $\overline{\omega}$, with the norm $|\cdot|_0$.

$C^k(\overline{\omega})$ is the space of k-times continuously differentiable functions defined in $\overline{\omega}$, with the norm $|\cdot|_k$.

$C^{k,\varepsilon}(\overline{\omega})$ (the Hölder space) is the space of functions from $C^k(\overline{\omega})$ whose kth derivatives satisfy the Hölder condition with exponent $\varepsilon \in [0,1]$, equipped with the norm $|\cdot|_{k,\varepsilon}$.

We introduce the spaces of functions defined on $\partial\omega$, Γ, or S in a smiilar way. We preserve the above notation for the norms in these spaces. If it is not clear from the context over which set the norm is taken, we indicate the set in the notation (e.g., $\|\cdot\|(\Gamma)$). For spaces of vector-valued functions we indicate the space to which the values of the functions belong in the notation (e.g., $C^{k,\varepsilon}(\overline{\omega},\mathbb{R}^n)$, $L_q(\partial\omega,\mathbb{R}^n)$). The norms in the spaces of vector-valued functions are denoted as above.

Necessary information about function spaces used in the book can be found in the Appendix. Throughout the book only real numbers are used.

CHAPTER 1

Linear Perturbations of the Operator div

Introduction

In this chapter we consider linear perturbations of the operator div and discuss properties of perturbated operators L. In §1, we study the existence of an analog of the operator rot for the operator L, i.e., the first order operator R such that $LR = 0$. We give examples of L for which nontrivial operators R do or do not exist. In §2, we study the existence of a nontrivial solution to the problem $L^*p = 0$ and prove several coercive estimates for the operator L^*. In §3, some well-known results about boundary-value problems for the operator div and some results on solenoidal vector fields are proved. We give the proofs in the form that will be convenient later. In §4, we consider a boundary-value problem for the operator $L = \operatorname{div}(\cdot) + (\sigma, \cdot)_m$ which is a compact perturbation of div. In §6, for the problem $Lu = 0$ in ω, $u\big|_{\partial\omega} = 0$ we construct some solutions, which allows us to prove that the kernel of L is infinite-dimensional, to derive conditions under which the quadratic integral forms on the kernel of L are nonnegative, and to study the question of reconstructing L from its kernel (cf. §7). In §8, we prove a theorem which indicates how to find vector fields from the kernel of L for a fixed projection, at each point x, onto a subspace depending on x. In the above problems, we are lloking for solutions in Hölder spaces. Estimates for the solution are established in the Hölder spaces, as well as in the Sobolev spaces. The main result in §9 is L_q-estimates for those components of a vector field f that appear in the decomposition of f into the sum of terms that are analogs of solenoidal and potential fields. In §10, we prove the Weyl decomposition of L and establish its stability.

§1. The operator L

Let $\omega \subset \mathbb{R}^m$ be a bounded domain, let $A(x) = \{A_{ij}(x)\}$, $i = 1, \dots, n$, $j = 1, \dots, m$, be a matrix-valued function, and let $a(x) = \{a^i(x)\}$, $i = 1, \dots, n$, be a vector-valued function. Introduce the matrix $G(x) = A^*(x)A(x)$ such that

$$\det G(x) \neq 0, \quad x \in \overline{\omega}. \tag{1.1}$$

Note that (1.1) is possible for an $m \times m$ matrix G only if $n \geq m$. Unless otherwise stated, we assume that

$$\partial\omega \in C^{k+1,\varepsilon}, \quad A_{ij} \in C^{k,\varepsilon}(\overline{\omega}), \quad a^i \in C^{k,\varepsilon}(\overline{\omega}) \tag{1.2}$$

for all i, j, some natural k, and $\varepsilon \in (0, 1]$. In this case, for every $x \in \overline{\omega}$ the operators

$$P = AG^{-1}A^*, \qquad Q = I - P \tag{1.3}$$

are the orthogonal projections onto complementary subspaces RA and NA^* in $\mathbb{R}^n$ and have matrix-valued coefficients of class $C^{k,\varepsilon}(\overline{\omega})$.

On the set of n-dimensional vector-valued functions, we define the operator L by the equality

$$Lu = \operatorname{div} A^*u + (a, u)_n. \tag{1.4}$$

LEMMA 1.1. *The coefficient $a(x)$ in (1.4) is uniquely represented in the form*

$$a(x) = A(x)h(x) + b(x), \tag{1.5}$$

where

$$\begin{gathered} b \in C^{k,\varepsilon}(\overline{\omega}, \mathbb{R}^n), \\ A^*(x)b(x) \equiv 0, \quad h(x) = \sigma(x) + \nabla\psi(x), \\ \sigma \in C^{k,\varepsilon}(\overline{\omega}, \mathbb{R}^m), \quad \psi \in C^{k+1,\varepsilon}(\overline{\omega}), \\ \operatorname{div}\sigma = 0, \quad \sigma_\nu\big|_{\partial\omega} = 0, \quad \int_\omega \psi(x)\,dx = 0. \end{gathered}$$

PROOF. The representation (1.5) holds because h and b are uniquely defined by the equalities $h = G^{-1}A^*a$ and $b = Qa$. The decomposition of h into the sum of the potential and solenoidal parts is also unique and can be realized by solving the problem

$$\Delta\psi = \operatorname{div} h, \quad \left(\frac{\partial\psi}{\partial\nu} - h_\nu\right)\bigg|_{\partial\omega} = 0, \quad \int_\omega \psi\,dx = 0, \quad \sigma = h - \nabla\psi.$$

□

LEMMA 1.2. *In $\overline{\omega}$, an $(n \times l)$-matrix-valued function D is uniquely represented in the form*

$$D = AG^{-1}D' + D'', \tag{1.6}$$

where $D' : \mathbb{R}^l \to \mathbb{R}^m$, $D'' : \mathbb{R}^l \to \mathbb{R}^n$, and $PD'' \equiv 0$.

PROOF. Indeed, D' and D'' can be defined by the equalities $D' = A^*D$ and $D'' = QD$. The uniqueness is obvious. □

The operator div is a special case of the operator L for $n = m$, $A \equiv I$, and $a \equiv 0$. The relation $\operatorname{div}\operatorname{rot} = 0$ for $n = m = 3$ is well known. The goal of this section is to describe all first order linear operators R that are defined on the set of l-dimensional vector-valued functions and satisfy the relation $LR = 0$, where the operator L is defined by (1.4).

Let $B_s : \mathbb{R}^l \to \mathbb{R}^n$, $s = 1, \dots, m$, and $C : \mathbb{R}^l \to \mathbb{R}^n$ be matrix-valued functions in $\overline{\omega}$. We introduce the operator R as follows:

$$R\varphi = B_s\varphi_{x_s} + C\varphi, \qquad \varphi \in C_0^\infty(\omega, \mathbb{R}^l). \tag{1.7}$$

Using Lemma 1.2 for the matrices B_s and C, we rewrite (1.7) in the form

$$(1.8)\quad \begin{gathered} u \equiv R\varphi \equiv AG^{-1}\left(B'_s\varphi_{x_s} + C'\varphi\right) + \left(B''_s\varphi_{x_s} + C''\varphi\right) \equiv AG^{-1}u' + u'', \\ B'_s : \mathbb{R}^l \to \mathbb{R}^m, \quad C' : r^l \to \mathbb{R}^m, \quad B''_s : \mathbb{R}^l \to \mathbb{R}^n, \quad C'' : \mathbb{R}^l \to \mathbb{R}^n, \\ A^*B''_s \equiv 0, \quad A^*C'' \equiv 0. \end{gathered}$$

LEMMA 1.3. *Let the entries of the matrices $B_s(x)$ and $C(x)$ be continuously differentiable in ω. Then the conditions*

$$(1.9)\quad \begin{gathered} B'_{sij} = -B'_{isj}, \\ (B'_{sij})_{x_i} + C'_{sj} + B'_{sij}h^i + B''_{sij}b^i = 0, \\ (C'_{ij})_{x_i} + C'_{ij}h^i + C''_{ij}b^i = 0 \end{gathered}$$

for all $x \in \omega$, where h and b are the functions in (1.5), *are necessary and sufficient for the equality $LR\varphi = 0$ to hold for any $\varphi \in C_0^\infty(\omega, \mathbb{R}^n)$.*

PROOF. By Lemma 1.2 for the matrices D and D', the matrices B'_s, C', B''_s, C'', $s = 1, \dots, m$, are continuously differentiable. Hence $u' \in C^1_{\text{loc}}(\omega, \mathbb{R}^m)$ and $u'' \in C^1_{\text{loc}}(\omega, \mathbb{R}^m)$. Since $A^*u'' \equiv 0$, we have $Lu = \operatorname{div} u' + (h, u')_m + (b, u'')_n$. Therefore,

$$(1.10)\quad \begin{aligned} LR\varphi = B'_{sij}\varphi^j_{x_s x_i} &+ \left[(B_{sij})_{x_i} + C'_{sj} + B'_{sij}h^i + B''_{sij}b^i\right]\varphi^j_{x_s} \\ &+ \left[\left(C'_{ij}\right)_{x_i} + C'_{ij}h^i + C''_{ij}b^i\right]\varphi^j. \end{aligned}$$

The sufficiency of (1.9) follows from (1.10). To prove the necessity, take $\varphi(x)$ of the special form

$$\begin{gathered} \varphi(x) = \zeta\alpha\left((x - x_0, \lambda)_m\right)\beta_\rho(x - x_0), \\ x_0 \in \omega, \quad \zeta \in \mathbb{R}^n, \quad \lambda \in \mathbb{R}^m, \end{gathered}$$

where $\alpha \in C^\infty(\mathbb{R}^1)$, $\beta_\rho \in C^\infty(\mathbb{R}^m)$, φ is identically equal to 1 if $|x - x_0| \le \rho/2$ and is equal to 0 if $|x - x_0| \ge \rho$, and the ball of center x_0 and radius ρ lies in ω. Since

$$\begin{gathered} \varphi^j(x_0) = \zeta_j\alpha(0), \qquad \varphi^j_{x_s}(x_0) = \zeta_j\lambda_s\alpha'(0), \\ \varphi^j_{x_s x_r}(x_0) = \zeta_j\lambda_s\lambda_r\alpha''(0), \end{gathered}$$

the necessity of (1.9) follows from the fact that the right-hand side of (1.10) vanishes for arbitrary $\alpha(0)$, $\alpha'(0)$, $\alpha''(0)$, ζ, λ, x_0. □

We reformulate Lemma 1.3 in a more convenient form. We introduce a skew-symmetric matrix $\xi(x)$ with entries $\xi_{ij} = h^i_{x_j} - h^j_{x_i}$, $i, j = 1, \dots, m$, and consider the Hilbert–Schmidt inner product $\langle B, D\rangle = \operatorname{tr} BD^*$ in the space of $m \times m$ matrices.

THEOREM 1.1. *The general form of the operator R with continuously differentiable coefficients such that $LR\varphi = 0$ for all $\varphi \in C_0^\infty(\omega, \mathbb{R}^l)$ is given by the formula*

$$(1.11)\quad R\varphi = AG^{-1}\varphi' + \varphi'',$$

where

$$\varphi'^r = \left(\varphi^j B'^j_{sr}\right)_{x_s} - \varphi^j \left(B'^j h + B''^j b\right)^r,$$
$$\varphi''^r = B''^j_{sr} \varphi^j_{x_s} + C''_{rj}\varphi^j,$$

where the matrix-valued functions $B'^j : \mathbb{R}^m \to \mathbb{R}^m$, $B''^j : \mathbb{R}^n \to \mathbb{R}^m$, *and* $C'' : \mathbb{R}^l \to \mathbb{R}^n$, $j = 1, \dots, l$, *satisfy the system*

$$\begin{gathered} B'^{j*} = -B'^j, \\ \tfrac{1}{2}\langle B'^j, \xi\rangle = (B''^j b, h)_m + \operatorname{div} B''^j b - (C''^* b)^j, \\ A^* C'' \equiv 0, \quad B''^j A \equiv 0, \quad j = 1, \dots, l; \quad \left(B'^j_{sr}\right)_{x_s} \in C^1_{\text{loc}}(\omega). \end{gathered} \tag{1.12}$$

PROOF. Let matrices B'^j and B''^j be defined as follows:

$$\begin{aligned} \left(B'^j\right)_{ir} &= B'_{irj}, \quad j = 1, \dots, l, \quad r = 1, \dots, m, \quad i = 1, \dots, m, \\ \left(B''^j\right)_{ir} &= B''_{irj}, \quad j = 1, \dots, l, \quad r = 1, \dots, n, \quad i = 1, \dots, m. \end{aligned}$$

It is easy to check that the system of equalities $A^* B''_j = 0$, $j = 1, \dots, m$, is equivalent to the system $B''^j A = 0$, $j = 1, \dots, l$. Then the last equalities in (1.12) are equivalent to the similar equalities in (1.8). It is obvious that the first equalities in (1.12) and in (1.9) are equivalent. The second formula in (1.9) implies $\left(B'_{sij}\right)_{x_i} \in C^1_{\text{loc}}(\omega)$, which is equivalent to the inclusion $\left(B'^j_{sr}\right)_{x_s} \in C^1_{\text{loc}}(\omega)$. To complete the proof of the theorem, it remains to exclude C'_{sj} from the second equality in (1.9) and substitute the result in the first equality in (1.8) and in the third equality in (1.9). The first substitution leads to (1.11), whereas the second one yields the second equality in (1.12). □

The system (1.12) for B'^j, B''^j, $j = 1, \dots, l$, and C'' is not necessarily solvable. To conclude the section, we consider two special cases of the operator L. In the first case, the system (1.12) does not admit a nontrivial solution whereas in the second case a nontrivial solution can exist.

EXAMPLE 1.1. Let $n = m$, $A \equiv I$, and $a \equiv h$. Then $Lu = \operatorname{div} u + (h, u)_m$. The last equalities in (1.12) yield $C'' \equiv 0$, $B''^j \equiv 0$, $j = 1, \dots, l$, and the equation for B'^j can be rewritten as follows:

$$\langle B'^j, \xi\rangle = 0, \quad B'^{j*} = -B'^j, \qquad j = 1, \dots, l. \tag{1.13}$$

Let $m = 2$ and $\xi(x) \neq 0$ in ω. Since the space of skew-symmetric 2×2 matrices is one-dimensional, any solution of (1.13) is zero. Thus, in the two-dimensional case, only the zero operator R exists for the above operator L.

EXAMPLE 1.2. Let $b \equiv 0$ and $h = \nabla\psi$ in (1.5). Then $\xi(x) \equiv 0$ in ω, and the second equality in (1.12) is valid. Therefore, the system (1.12) takes the form

$$\begin{gathered} B'^{j*} = -B'^j, \quad B''^j A = 0, \quad j = 1, \dots, l, \\ A^* C'' = 0, \quad \left(B'^j_{sr}\right)_{x_s} \in C^1_{\text{loc}}(\omega). \end{gathered} \tag{1.14}$$

It is obvious that (1.14) admits a nontrivial solution, e.g., $C'' \equiv 0$, $B''^j \equiv 0$ and B'^j are nonzero smooth skew-symmetric matrices. We note that the representation of φ'^r mentioned in Theorem 1.1 can be simplified as follows:

$$\varphi'^r = \left(\varphi^j B'^j_{sr}\right)_{x_s} - \varphi^j \left(B'^j_{rs}\psi_{x_s}\right) = e^{-\psi}\left[\varphi^j e^{\psi} B'^j_{sr}\right]_{x_s}.$$

If we set $n = m$, $A \equiv I$, and $\psi \equiv 0$ in these equalities, then (1.14) implies $B''^j = 0$ and $C'' = 0$. Consequently, the general form of the operator R for $L = \operatorname{div}$ is given by

$$(R\varphi)^r = \left(\varphi^j B'^j_{sr}\right)_{x_s}, \quad r = 1, \dots, m.$$

§2. The operator L^*

Denote by L^* the operator formally adjoint to L. It is defined on the set of scalar functions by the formula

$$L^* p = -A\nabla p + ap. \tag{2.1}$$

We find conditions under which the problem

$$L^* p = 0, \quad p \in C^1(\overline{\omega}), \tag{2.2}$$

has a nontrivial solution.

THEOREM 2.1. *The problem* (2.2) *has a nontrivial solution if and only if* $b \equiv 0$ *and* $\sigma \equiv 0$ *in* (1.5). *In this case, any solution to* (2.2) *is proportional to* $p^* = e^{\psi}$.

PROOF. *Sufficiency.* If $b \equiv 0$ and $\sigma \equiv 0$, then $L^*p = -A\nabla p + pA\nabla\psi = -e^{\psi}[A\nabla(e^{-\psi}p)]$. Since the matrix $G(x)$ is nonsingular, problem (2.2) is equivalent to the equation $\nabla e^{-\psi}p = 0$, whence $p = ce^{\psi}$.

Necessity. Using (1.5), we rewrite (2.2) in the form $A[\nabla p - (\nabla\psi + \sigma)p] = bp$. Since the matrix G is invertible, the equality $A^*b \equiv 0$ implies that the above equation is equivalent to the system

$$\nabla p = (\nabla\psi + \sigma)p, \quad bp = 0. \tag{2.3}$$

We assume that (2.3) has a nontrivial solution $p \in C^1(\overline{\omega})$ and prove that p does not vanish at any point of $\overline{\omega}$. Let $x(t)$ be a continuously differentiable curve contained in $\overline{\omega}$. Along x, a solution to the first equation of (2.3) satisfies the following linear first order ordinary differential equation:

$$\frac{d}{dt} p(x(t)) = (\nabla\psi(x(t)) + \sigma(x(t)), \dot{x}(t))_m \, p(x(t)).$$

Therefore, either $p(x(t))$ is identically zero or the sign of $p(x(t))$ is constant along the entire curve $x(t)$. Since any two points of $\overline{\omega}$ can be joined by such a curve, the sign of a nontrivial solution to the first equation of the system (2.3) is constant in $\overline{\omega}$. Hence $\ln|p(x)| \in C^1(\overline{\omega})$ and (2.3) imply $\nabla(\ln|p| - \psi) = \sigma$. Since σ is orthogonal to the gradient of any continuously differentiable function of $L_2(\omega, \mathbb{R}^m)$, the last equality holds only if $\sigma \equiv 0$ and $p = ce^{\psi}$. The second equality of the system (2.3) means that $b \equiv 0$. □

In Theorem 2.1, a solution to the problem (2.2) is assumed to be continuously differentiable. We show that it is possible to weaken this condition on the smoothness of the solution.

LEMMA 2.1. *Let $p \in W_r^1(\omega)$, $r \geq 1$, be a solution to the equation $L^*p = f$, where $f \in C^{k',\varepsilon'}(\overline{\omega}, \mathbb{R}^n)$, $k' \geq 0$, $\varepsilon' \in [0,1]$. Then $p(x)$ coincides up to a set of measure zero with a function of class $C^{k''+1,\varepsilon''}(\overline{\omega})$, where $k'' = \min\{k, k'\}$ and $\varepsilon'' = \min\{\varepsilon, \varepsilon'\}$.*

PROOF. From the equality $L^*p = f$ it follows that

$$\nabla p = pG^{-1}A^*a - G^{-1}A^*f. \tag{2.4}$$

Since $W_r^1(\omega)$ is embedded into $L_{r'}(\omega)$, the right-hand side of (2.4) belongs to the space $L_{r'}(\omega, \mathbb{R}^m)$, where $r' = mr/(m-r)$ if $m > r$ and r' is arbitrary if $r \geq m$. Hence (2.4) yields the inclusion $p \in W_{r'}^1(\omega)$. Repeating the procedure s times, we find that $p \in W_q^1(\omega)$, where $q = mr/(m - sr)$ if $m > sr$ and q is arbitrary if $m \leq sr$. For sufficiently large s we have $m \leq sr$. Since $W_q^1(\omega)$, $q > m$, is embedded into $C(\overline{\omega})$, the right-hand side of (2.4) is continuous (perhaps, after a correction of $p(x)$ on a set of measure zero). Therefore, $p \in C^1(\overline{\omega})$. Repeating the procedure k'' times, we finally find that $p \in C^{k''+1}(\overline{\omega})$. In this case, the right-hand side of (2.4) belongs to $C^{k'',\varepsilon'}(\overline{\omega}, \mathbb{R}^n)$, which implies the required smoothness of p. □

To conclude the section, we give a number of coercive estimates for L.

LEMMA 2.2. *For all $q \in [1, \infty)$, $s = 0, \dots, k$, $\mu \in [0, \varepsilon]$ a function $p \in C^{k+1,\varepsilon}(\overline{\omega})$ satisfies the following estimates:*

$$\begin{aligned} \|p\|_{1,q} &\leq C_q\left(\|L^*p\|_q + \|p\|_1\right), \\ |p|_{s+1,\mu} &\leq C_s\left(|L^*p\|_{s,\mu} + \|p\|_1\right). \end{aligned} \tag{2.5}$$

PROOF. Let $L^*p = f$. Then (2.4) holds. Taking the L_q-norm of both sides of (2.4), we obtain

$$\|p\|_{1,q} \leq C\left(\|p\|_q + \|f\|_q\right). \tag{2.6}$$

Since $W_r^1(\omega)$ is embedded into $L_q(\omega)$, it follows that $\|p\|_q \leq C\|p\|_{1,r}$, where $r = \max\{qm/(q+m), 1\}$. Combining the last inequality with (2.6), we get

$$\|p\|_{1,q} \leq C\left(\|p\|_{1,r} + \|f\|_q\right). \tag{2.7}$$

To estimate $\|p\|_{1,r}$ from above, we use (2.6) with q replaced by r. Since $r < q$, (2.7) takes the form

$$\|p\|_{1,q} \leq C\left(\|p\|_r + \|f\|_q\right). \tag{2.8}$$

Iterating the above procedure s times, we establish the validity of (2.8) with $r = \max\{qm/(sq+m), 1\}$. If $s \geq m$, then $r = 1$. In this case, (2.8) coincides with the first inequality in (2.5). Taking the $C^{s,\mu}$-norm of both sides of (2.4), we get

$$|p|_{s+1,\mu} \leq C\left(|p|_{s,\mu} + |f|_{s,\mu}\right).$$

Iterating this inequality s times, we find that

$$|p|_{s+1,\mu} \leq C\left(|p|_{0,\mu} + |f|_{s,\mu}\right). \tag{2.9}$$

For sufficiently large q and for $\mu < 1$ the space $W_q^1(\omega)$ is embedded into the space $C^{0,\mu}(\overline{\omega})$, and the estimate $|p|_{0,\mu} \leq C\|p\|_{1,q}$ holds. From (2.9), the last estimate, and the first estimate in (2.5) which is already proved, we obtain the second estimate in (2.5). □

LEMMA 2.3. *Suppose that the problem* (2.2) *has only the zero solution. Then for* $q \in (1, \infty)$, $s = 0, \dots, k$, $\mu \in [0, \varepsilon]$ *a function* $p \in C^{k+1,\varepsilon}(\overline{\omega})$ *satisfies the following estimates*:

$$\|p\|_{1,q} \le C_q \|L^* p\|_q, \quad |p|_{s+1,\mu} \le C_s |L^* p|_{s,\mu}. \tag{2.10}$$

PROOF. By Lemma 2.2, to prove (2.10) it suffices to obtain the estimate

$$\|p\|_q \le C_q \|L^* p\|_q \tag{2.11}$$

for $p \in C^{k+1,\varepsilon}(\overline{\omega})$. We prove (2.11) by contradiction. We assume that there exists a sequence $p_n \in C^{k+1,\varepsilon}(\overline{\omega})$ such that

$$\|p_n\|_q = 1, \quad \|L^* p_n\|_q \to 0. \tag{2.12}$$

By (2.4), the norms $\|p_n\|_{1,q}$ are uniformly bounded. Since the space $W_q^1(\omega)$ is reflexive for $q > 1$, we can find a subsequence of p_n (we keep the same notation p_n) that converges weakly to $p \in W_q^1(\omega)$ in $W_q^1(\omega)$. By the compactness of the embedding of $W_q^1(\omega)$ into $L_q(\omega)$ and formulas (2.12) and (2.4) for $\nabla p_n(x)$, the subsequence p_n strongly converges to p. Passing to the limit in (2.12), we obtain

$$\|p\|_q = 1, \quad L^* p = 0, \quad p \in W_q^1(\omega). \tag{2.13}$$

By Lemma 2.1 with $f \equiv 0$, the function p is continuously differentiable. Thus, p is a nonzero solution to problem (2.2), contradicting the above assumption. □

LEMMA 2.4. *Let the problem* (2.2) *have a nonzero solution. Then for* $s = 0, \dots, k$, $q \in (1, \infty)$, $\mu \in [0, \varepsilon]$ *a function* $p \in C^{k+1,\varepsilon}(\overline{\omega})$ *satisfies the estimates*

$$\begin{aligned} \|p\|_{1,q} &\le C_q \left[\|L^* p\|_q + \Phi(p)\right], \\ |p|_{s+1,\mu} &\le C_s [|L^* p|_{s,\mu} + \Phi(p)], \end{aligned} \tag{2.14}$$

where $\Phi(p)$ *is a seminorm which is bounded in* $W_q^1(\omega)$ *for each* $q \in (1, \infty)$ *and satisfies* $\Phi(p^*) \ne 0$.

PROOF. By Lemma 2.2, to prove (2.14) it suffices to obtain the estimate

$$\|p\|_q \le C_q \left[\|L^* p\|_q + \Phi(p)\right], \quad 1 < q < \infty, \tag{2.15}$$

for $p \in C^{k+1,\varepsilon}(\overline{\omega})$. We prove (2.15) by contradiction. We assume that there exists a sequence $p_n \in C^{k+1,\varepsilon}(\overline{\omega})$ such that

$$\|p_n\|_q = 1, \quad \|L^* p_n\|_q \to 0, \quad \Phi(p_n) \to 0. \tag{2.16}$$

As in Lemma 2.3, we find a subsequence (denoted by p_n) that strongly converges to $p \in W_q^1(\omega)$ in $W_q^1(\omega)$. Passing to the limit in (2.16), we get

$$\|p\|_q = 1, \quad L^* p = 0, \quad \Phi(p) = 0, \quad p \in W_q^1(\omega). \tag{2.17}$$

Hence $p \in C^1(\overline{\omega})$, i.e., p is a solution to the problem (2.2). By Theorem 1.1, $p(x) = \alpha p^*(x)$ for some α. Since $\Phi(p^*) \ne 0$, the last equality in (2.17) implies $\alpha = 0$. Consequently, $p \equiv 0$, which contradicts the first equality in (2.17). □

§3. The operator div

In this section, we study properties of boundary-value problems for the operator div. We begin with the following problem:

$$\operatorname{div} w = 0, \quad w\big|_{\partial\omega} = \chi; \quad \chi \in C^{k,\varepsilon}(\partial\omega, \mathbb{R}^m), \quad \chi_\nu\big|_{\partial\omega} = 0. \tag{3.1}$$

LEMMA 3.1. *Let $\omega \subset \mathbb{R}^m$ be a bounded domain with boundary $\partial\omega \in C^{k+1,\varepsilon}$, where k is a natural number and $\varepsilon \in [0,1]$. There exists a solution $w \in C^{k,\varepsilon}(\overline{\omega}, \mathbb{R}^m)$ to the problem (3.1) that depends linearly on χ and satisfies the estimates*

$$|w|_{s,\varepsilon} \le C_s|\chi|_{s,\varepsilon}, \quad \|w\|_{1,q} \le C_q\|\chi\|_{1-1/q,q}, \tag{3.2}$$

where $s = 1, \ldots, k$ and $q \in (1, \infty)$.

PROOF. We seek a solution to the equation $\operatorname{div} w = 0$ in the form $w^r = (R\varphi)^r = (\varphi^j B'^j_{tr})_{x_t}$, $r = 1, \ldots, m$, where B'^j, $j = 1, \ldots, l$, are skew-symmetric matrices (cf. Example 1.2). Let us choose matrices B'^j and a vector-valued function φ so as to satisfy all the assumptions of the lemma.

Let $d(x)$ be the oriented distance from $x \in \mathbb{R}^m$ to $\partial\omega$ (cf. the Appendix). The distance $d(x)$ is defined in a sufficiently narrow strip and belongs to the space $C^{k+1,\varepsilon}$ there. We fix a function $\alpha \in C_0^\infty(\mathbb{R}^1)$ such that $\alpha(t) \equiv 0$ for $|t| \ge \delta$ and $\alpha(t) \equiv t$ for $|t| \le \delta/2$ with δ small enough. We also define a vector-valued function φ and matrices B'^j by the equalities

$$\varphi^j(x) = \zeta_j \alpha(d(x)), \quad B'^j(x) = \zeta_j B'(x),$$

where $B'(x)$ is a skew-symmetric matrix and $\zeta \in \mathbb{R}^l$ is a unit vector. Let Π be the extension operator from $\partial\omega$ (cf. the Appendix). We introduce vector-valued functions $\widehat{\chi}(x)$ and $\widehat{\nu}(x)$, $x \in \mathbb{R}^m$, as follows:

$$\widehat{\chi}(x) = (\Pi\chi)(x), \quad \widehat{\nu}(x) = (\Pi\nu)(x).$$

For $B'(x)$ we take the skew-symmetric matrix

$$B'(x) = \widehat{\nu}(x)\,(\widehat{\chi}(x), \cdot)_m - \widehat{\chi}(x)\,(\widehat{\nu}(x), \cdot)_m\,.$$

Since $B'_{tr} = \widehat{\chi}^r\widehat{\nu}^t - \widehat{\chi}^t\widehat{\nu}^r$, we obtain

$$\begin{aligned} w^r &= \left\{\alpha(d)\left(\widehat{\nu}^t\widehat{\chi}^r - \widehat{\nu}^r\widehat{\chi}^t\right)\right\}_{x_t} = \alpha'(d)d_{x_t}\left[\widehat{\nu}^t\widehat{\chi}^r - \widehat{\nu}^r\widehat{\chi}^t\right] + \left[\alpha(d)\widehat{\nu}^t_{x_t}\right]\widehat{\chi}^r \\ &\quad + \left[\alpha(d)\widehat{\chi}^r_{x_t}\right]\widehat{\nu}^t - \left[\alpha(d)\widehat{\nu}^r_{x_t}\right]\widehat{\chi}^t - \left[\alpha(d)\widehat{\chi}^t_{x_t}\right]\widehat{\nu}^r, \qquad r = 1, \ldots, m. \end{aligned}$$

Using the properties of the operator Π, we can derive (3.1) and show that w depends linearly on χ. Since $d(x) = 0$ and $\nabla d(x) = \nu(x)$ for $x \in \partial\omega$, the boundary condition

$$w(x) = \chi(x) - \nu(x)\,(\chi(x), \nu(x))_m = \chi(x)$$

holds on $\partial\omega$. □

We study the following more general problem for the operator div:

$$\operatorname{div} v = f, \quad v\big|_{\partial\omega} = \varphi, \quad f \in C^{k-1,\varepsilon}(\overline{\omega}), \quad \varphi \in C^{k,\varepsilon}(\partial\omega, \mathbb{R}^m). \tag{3.3}$$

THEOREM 3.1. *Let $\omega \subset \mathbb{R}^m$ be a bounded domain with boundary $\partial\omega \in C^{k+1,\varepsilon}$, where k is a natural number and $\varepsilon \in (0,1]$. A necessary and sufficient condition for the existence of a solution $v \in C^{k,\varepsilon}(\overline{\omega}, \mathbb{R}^m)$ to the problem* (3.3) *is the following consistency condition*:

$$\int_\omega f\,dx = \int_{\partial\omega} \varphi_\nu\,dS. \tag{3.4}$$

If (3.4) *holds, then there exists a solution $v \in C^{k,\varepsilon}(\overline{\omega}, \mathbb{R}^m)$ to the problem* (3.3) *that depends linearly on f, φ and satisfies the estimates*

$$\begin{aligned} |v|_{s,\varepsilon} &\le C_s\left(|\varphi|_{s,\varepsilon} + |f|_{s-1,\varepsilon}\right), \\ \|v\|_{1,q} &\le C_q\left(\|\varphi\|_{1-1/q,q} + \|f\|_q\right) \end{aligned} \tag{3.5}$$

for $s = 1, \dots, k$ and $q \in (1, \infty)$.

PROOF. The necessity of the consistency condition (3.4) can be obtained by integrating both sides of the equation and using the integration by parts formula and the boundary condition.

Let (3.4) hold. We seek a solution to the problem (3.3) in the form $v = \nabla p + w$, where p and w are determined from the relations

$$\Delta p = f, \quad \left.\frac{\partial p}{\partial \nu}\right|_{\partial\omega} = \varphi_\nu, \quad \int_\omega p\,dx = 0,$$
$$\operatorname{div} w = 0, \quad w\big|_{\partial\omega} = \varphi - \nabla p\big|_{\partial\omega}.$$

With respect to p this problem is uniquely solvable. Its solution p depends linearly on f and φ, belongs to the space $C^{k+1,\varepsilon}(\overline{\omega})$, and satisfies the estimates

$$\begin{aligned} |p|_{s+1,\varepsilon} &\le C_s\left(|\varphi|_{s,\varepsilon} + |f|_{s-1,\varepsilon}\right), \\ \|p\|_{2,q} &\le C_q\left(\|\varphi\|_{1-1/q,q} + \|f\|_q\right), \end{aligned}$$

where $s = 1, \dots, k$ and $q \in (1, \infty)$. Since $\left(\varphi \nabla p\big|_{\partial\omega}\right) = 0$, for w we can take the solution from Lemma 3.1. Then the function v satisfies all the requirements of the theorem. □

From the explicit construction of the operator R it follows that there exist smooth solenoidal vector-valued functions different from zero. The question arises: what part of the space of all vector-valued functions do such functions fill? The answer is given by Theorem 3.2. We begin the proof with an auxiliary lemma.

LEMMA 3.2. (a) *Let $u \in C^{k,\varepsilon}(\overline{\omega}, \mathbb{R}^m)$, $\operatorname{div} u = 0$, $u_\nu\big|_{\partial\omega} = 0$, and $\int_\omega (u, v)_m = 0$ for vector-valued functions $v \in C^{k,\varepsilon}(\overline{\omega}, \mathbb{R}^m)$ such that $\operatorname{div} v = 0$ and $v_{\partial\omega} = 0$. Then $u \equiv 0$.*

(b) *Let a function $w \in C^{k,\varepsilon}(\overline{\omega}, \mathbb{R}^m)$ satisfy the equality $\int_\omega (w, v)_m\,dx = 0$ for the functions $v(x)$ mentioned in* (a). *There exists $p \in C^{k+1,\varepsilon}(\overline{\omega})$ such that $w(x) = \nabla p(x)$.*

PROOF. (a) Let $\varphi \in C^\infty(\mathbb{R}^1_+)$, $\varphi(t) = t$ for $t \le 1/2$, $\varphi(t) \equiv 1$ for $t \ge 1$, and $0 < \varphi(t) \le 1$ for $t > 0$. For a sufficiently small δ we define $\chi_\delta \in C^{k+1,\varepsilon}(\overline{\omega})$ by the equality $\chi_\delta(x) = \varphi(\delta^{-1}\rho(x))$, where $\rho(x)$ is the distance from $x \in \overline{\omega}$ to $\partial\omega$ (cf. the Appendix). It is clear that $\chi_\delta(x)$ differs from 1 only in the boundary strip Γ_δ of width δ. Let $\widehat{u}(x) = u(x)\chi_\delta(x) \in C^{k,\varepsilon}(\overline{\omega}, \mathbb{R}^m)$. Then $\operatorname{div}\widehat{u} = (u, \nabla\chi_\delta)_m \in C^{k,\varepsilon}(\overline{\omega}, \mathbb{R}^m)$ and $\widehat{u}\big|_{\partial\omega=0}$. Since $\nabla\chi_\delta(x) = \delta^{-1}\dot{\varphi}(\delta^{-1}\rho(x))\nabla\rho(x)$ and $\nabla\rho(x) = -\nu(x)$ on $\partial\omega$, we have $(u, \nabla\rho) = 0$ on $\partial\omega$. Since $(u(x), \nabla\rho(x))_m$ as a function of x belongs to $C^{k,\varepsilon}(\Gamma_\delta)$, $k \ge 1$, the function $(u, \nabla\rho)_m$ it is at least continuously differentiable. Hence

$$|(u(x), \nabla\rho(x))_m| \le C\rho(x) \le C\delta, \qquad x \in \Gamma_\delta.$$

Thus,

$$|\operatorname{div}\widehat{u}| \le C, \quad x \in \Gamma_\delta, \qquad \operatorname{div}\widehat{u}(x) = 0, \quad x \in \omega\backslash\Gamma_\delta,$$

which implies

$$\|\operatorname{div}\widehat{u}(x)\|_q \le C\delta^{1/q} \to 0 \quad \text{as } \delta \to 0 \text{ for } q \in [1, \infty).$$

Let $w(x)$ be the solution to the problem

$$\operatorname{div} w = \operatorname{div}\widehat{u}, \qquad w\big|_{\partial\omega} = 0$$

from Lemma 3.1. For $q > 1$ we have

$$\|w\|_{1,q} \le C_q \|\operatorname{div}\widehat{u}\|_q \to 0 \qquad \text{as} \quad \delta \to 0,$$

and the function $v = \widehat{u} - w \in C^{k,\varepsilon}(\overline{\omega}, \mathbb{R}^m)$ satisfies $\operatorname{div} v = 0$ and $v\big|_{\partial\omega} = 0$. Since the integral of $(u, v)_m$ over ω vanishes, we get

$$\int\limits_{\omega\backslash\Gamma_\delta} |u(x)|^2\,dx \le \int\limits_\omega \chi_\delta(x)|u(x)|^2\,dx \le \int\limits_\omega |(u, w)_m|\,dx \le \|u\|_2\|w\|_2. \tag{3.6}$$

We fix $q > 1$ so that the embedding operator of $W_q^1(\omega)$ into $L_2(\omega)$ is bounded. The left-hand side of (3.6) tends to zero as $\delta \to 0$, which is possible only if $u \equiv 0$.

(b) We represent w in the form $w = u + \nabla p$, where u and p are determined from the relations

$$\Delta p = \operatorname{div} w, \quad \frac{\partial p}{\partial\nu}\bigg|_{\partial\omega} = w_\nu, \quad \int\limits_\omega p\,dx = 0, \quad u = w - \nabla p.$$

Then $\operatorname{div} u = 0$ and $u_\nu\big|_{\partial\omega} = 0$. Using the assumptions of the lemma and integrating by parts, we obtain

$$0 = \int\limits_\omega (w, u)_m\,dx = \int\limits_\omega (u, v)_m\,dx.$$

In view of (a), we have $u \equiv 0$; therefore, $w = \nabla p$. □

Lemma 3.2 is used in the proof of the following theorem.

THEOREM 3.2. *Let ω be a bounded domain with boundary $\partial\omega \in C^1$ and let $u \in L_q(\omega, \mathbb{R}^m)$, $q \ge 1$. If $\int_\omega (u, v)_m\,dx = 0$ for any smooth solenoidal vector-valued function v with compact support, then there exists a function $p \in W_q^1(\omega)$ such that $u = \nabla p$.*

PROOF. Let a domain ω' be contained in the interior of ω and let $\partial\omega' \in C^{k+1,\varepsilon}$, $k \geq 1$, $\varepsilon \in (0,1]$. We define

$$w \in C^{k,\varepsilon}(\overline{\omega}', \mathbb{R}^m), \qquad \operatorname{div} w(x) = 0, \quad x \in \omega',$$
$$w\big|_{\partial\omega'} = 0, \qquad w(x) \equiv 0, \quad x \in \omega \backslash \omega'.$$

The average $w_\rho(x)$ of $w(x)$ over $0 < \rho < \operatorname{dist}(\partial\omega', \partial\omega)$ is a smooth solenoidal vector-valued function with compact support. By the assumptions of the theorem,

$$0 = \int_\omega (u, w_\rho)_m \, dx = \int_\omega (u_\rho, w)_m \, dx. \tag{3.7}$$

From (3.7) and Lemma 3.2(b) it follows that there exists a function $p(x,\rho) \in C^{k+1,\varepsilon}(\overline{\omega}')$ such that $u_\rho(x) = \nabla p(x,\rho)$ for $x \in \overline{\omega}'$. Since the function $p(x,\rho)$ is defined up to a constant, we can assume that the relation

$$\int_{\omega_0} p(x,\rho)\, dx = 0$$

holds in some subdomain $\omega_0 \subset \overline{\omega}_0 \subset \omega'$. Let $\rho_l > 0$ and $\rho_l \to 0$ as $l \to \infty$. By the Poincaré inequality, for sufficiently large r and s we have

$$\|p(x,\rho_r) - p(x,\rho_s)\|_q (\omega') \leq C \left\|\nabla \left(p(x,\rho_r) - p(x,\rho_s)\right)\right\|_q (\omega'),$$

where $C = C(\omega', \omega_0, q)$. Since $\nabla p(x,\rho_s) = u_{\rho_s}(x) \to u(x)$ in $L_q(\omega')$ as $s \to \infty$, the right-hand side of the last inequality tends to zero as $r, s \to \infty$. Therefore, $p(x,\rho_s)$ is a Cauchy sequence in $L_q(\omega')$, and its limit $p \in W_q^1(\omega')$ satisfies the equality

$$\nabla p(x) = u(x), \quad x \in \omega'.$$

Let ω_s, $s = 1, 2, \ldots$, be an increasing sequence of domains with boundaries $\partial\omega_s \in C^{k+1,\varepsilon}$, $k \geq 1$, $\varepsilon \in (0,1]$, such that $\overline{\omega}_0 \subset \omega_s \subset \overline{\omega}_s \subset \omega$ and every point $x \in \omega$ gets into ω_s for sufficiently large s. Hence, in every subdomain ω_s, a function $p(x,s)$ of class $W_q^1(\omega_s)$ can be defined such that

$$\nabla p(x,s) = u(x), \quad x \in \omega_s, \qquad \int_{\omega_0} p(x,s)\, dx = 0.$$

If $r > s$, then the application of the Poincaré inequality to $p(x,r) - p(x,s)$ in ω_s shows that $p(x,r)$ and $p(x,s)$ coincide almost everywhere in ω_s. Thus, for almost every $x \in \omega$ the quantity $p(x,r)$, starting from some number r, remains constant. We denote it by $p(x)$. It is obvious that $p(x)$ satisfies the relations

$$p \in L_{q,\mathrm{loc}}(\omega), \quad \nabla p = u \in L_q(\omega), \qquad \int_{\omega_0} p(x)\, dx = 0.$$

To complete the proof of the theorem, it remains to establish the inclusion $p \in L_q(\omega)$, i.e., to show that the function $p(x)$ is q-integrable in a sufficiently small neighborhood of every point of the boundary $\partial\omega$.

Let a neighborhood V of a point $x_0 \in \partial\omega$ have such a form and be so small that $V_0 = V \cap \omega$ satisfies the following conditions:

$$\partial V_0 \in C^1,$$
$$\text{there exists } \lambda \in \mathbb{R}^m \text{ such that } |\lambda| = 1,$$
$$\{\overline{V_0 + t\lambda}\} \subset \omega \text{ for sufficiently small } t > 0.$$

Such a neighborhood V exists because $\partial\omega$ is continuously differentiable. Consider a subdomain $V_1 \subset \overline{V}_1 \subset V_0$. By the Poincaré inequality,

$$\left\| p - |V_1|^{-1} \int\limits_{V_1 + t\lambda} p\, dx \right\|_q (V_0 + t\lambda) \leq C \|\nabla p\|_q (V_0 + t\lambda) \leq C\|u\|_q(\omega),$$

where $C \neq C(t)$. As t tends to zero, we arrive at the estimate

$$\left\| p - |V_1|^{-1} \int\limits_{V_1} p\, dx \right\|_q (V_0) \leq \|u\|_q(\omega),$$

whence $\|p\|_q(V_0)$ is finite. □

For $q = 2$ Theorem 3.2 describes the orthogonal complement to the linear set of smooth solenoidal functions with compact support in $L_2(\omega, \mathbb{R}^m)$, and is called the Weyl decomposition.

For future use, we need to modify the solution to the problem (3.3) from Theorem 3.1 so as to obtain a solution possessing some additional properties.

LEMMA 3.3. *Let the assumptions of Theorem* 3.1 *hold and let* $\sigma \in C^1(\overline{\omega}, \mathbb{R}^m)$, $\operatorname{div}\sigma = 0$, $\sigma_\nu\big|_{\partial\omega} = 0$, $\sigma \not\equiv 0$. *The solution* v *to the problem* (3.3) *from Theorem* 3.1 *possesses the additional property that*

$$\int_\omega (\sigma, v)_m \, dx = 0.$$

PROOF. We show that there exists a smooth solenoidal vector-valued function $\gamma_0(x)$ with compact support such that the integral of $(\gamma_0, \sigma)_m$ over ω is different from zero. Indeed, if for all such functions γ_0 the integral of $(\gamma_0, \sigma)_m$ vanishes, then $\sigma(x) = \nabla p(x)$ by Theorem 3.2. Since $\sigma \in C^1(\overline{\omega}, \mathbb{R}^m)$, the function p belongs to $C^2(\overline{\omega})$ and satisfies the relations

$$\Delta p = 0, \qquad \left.\frac{\partial p}{\partial \nu}\right|_{\partial\omega} = 0,$$

which follows from the properties of σ. In this case, $p \equiv \text{const}$ and $\sigma \equiv 0$. If v is the solution to the problem (3.3) from Theorem 3.1, then the function

$$v = \gamma_0 \int\limits_\omega (v, \sigma)_m \, dx \left[\int\limits_\omega (\gamma_0, \sigma)_m \, dx \right]^{-1}$$

satisfies all the requirements of the lemma. □

§4. Perturbation of the operator div

In this section, we study a boundary-value problem for the operator L of the form (1.4) with $n = m$, $A = I$, $a \equiv \sigma \not\equiv 0$, i.e.,

$$Lu = \operatorname{div} u + (\sigma, u)_m. \tag{4.1}$$

We recall that σ satisfies (1.5). As was shown in Example 1.1, it may occur that there exists no nontrivial operator R for the operator L defined above. Therefore, the method described in §3 is not applicable to this problem.

We begin with the following problem for an operator L of the form (4.1):

$$Lv = g, \quad v\big|_{\partial\omega} = 0, \quad g \in C^{k-1,\varepsilon}, \quad \int_\omega g(x)\,dx = 0. \tag{4.2}$$

LEMMA 4.1. *Let $\omega \subset \mathbb{R}^m$ be a bounded domain with boundary $\partial\omega \in C^{k+1,\varepsilon}$, where k is a natural number and $\varepsilon \in (0, 1]$. There exists a solution $v \in C^{k,\varepsilon}(\overline{\omega}, \mathbb{R}^m)$ to problem (4.2) that depends linearly on $g(x)$, is orthogonal to $\sigma(x)$ in $L_2(\omega, \mathbb{R}^m)$, and satisfies the estimates*

$$|v|_{s,\varepsilon} \le C_s |g|_{s-1,\varepsilon}, \qquad \|v\|_{1,q} \le C_q \|g\|_q \tag{4.3}$$

for $s = 1, \dots, k$ and $q \in (1, \infty)$.

PROOF. We introduce the space of vector-valued functions

$$\widehat{C}^{k,\varepsilon}(\overline{\omega}, \mathbb{R}^m) = \{u \in C^{k,\varepsilon}(\overline{\omega}, \mathbb{R}^m) : \int_\omega (\sigma, u)_m\,dx = 0\}$$

and denote by

$$\widehat{L}_q(\omega, \mathbb{R}^m), \quad 1 < q < \infty, \qquad \widehat{C}^{s-1,\varepsilon}(\overline{\omega}, \mathbb{R}^m), \quad s = 1, \dots, k, \tag{4.4}$$

the scales of spaces obtained by completing $\widehat{C}^{k,\varepsilon}(\overline{\omega}, \mathbb{R}^m)$ in the $L_q(\omega, \mathbb{R}^m)$-norm and $C^{s-1,\varepsilon}(\overline{\omega}, \mathbb{R}^m)$-norm respectively.

The space $\widehat{L}_q(\omega, \mathbb{R}^m)$ consists of all functions $u \in L_q(\omega, \mathbb{R}^m)$ such that the integral of $(\sigma, u)_m$ over ω vanishes, and $\widehat{C}^{s-1,\varepsilon}(\overline{\omega}, \mathbb{R}^m)$ is the subspace of $C^{s-1,\varepsilon}(\overline{\omega}, \mathbb{R}^m)$ formed by the functions in $C^{s-1,\varepsilon}(\overline{\omega}, \mathbb{R}^m)$ that satisfy the same orthogonality condition (cf. the Appendix).

Let $v \in \widehat{C}^{k,\varepsilon}(\overline{\omega}, \mathbb{R}^m)$, let g denote the right-hand side of the equation in (4.2), and let Bv and w be solutions from Lemma 3.3 to the problems

$$\operatorname{div} Bv = (\sigma, v)_m, \quad Bv\big|_{\partial\omega} = 0, \quad Bv \in \widehat{C}^{k,\varepsilon}(\overline{\omega}, \mathbb{R}^m), \tag{4.5}$$

$$\operatorname{div} w = g, \quad w\big|_{\partial\omega} = 0, \quad w \in \widehat{C}^{k,\varepsilon}(\overline{\omega}, \mathbb{R}^m), \tag{4.6}$$

respectively. We note that the smoothness of Bv is determined by the smoothness of the boundary $\partial\omega$. The solution Bv depends linearly on v and satisfies the estimates

$$|Bv|_{s,\varepsilon} \le C_s |v|_{s-1,\varepsilon}, \qquad \|Bv\|_{1,q} \le C_q \|v\|_q$$

for $s = 1, \dots, k$ and $q \in (1, \infty)$. Therefore, the operator B can be extended by continuity to a compact linear operator acting on each space of the scales (4.4). We use the same notation B for all these extensions. With the help of the operator B

and the function w, we can rewrite the problem (4.2) in the following (equivalent) form:

$$v + Bv + \gamma = w, \tag{4.7}$$

where $\gamma = \gamma(x)$ is a function such that

$$\gamma \in \widehat{C}^{k,\varepsilon}(\overline{\omega}, \mathbb{R}^m), \quad \gamma\big|_{\partial\omega} = 0, \quad \operatorname{div}\gamma = 0.$$

Let γ_j, $j = 1, \dots, r$, be a finite collection of such functions and let l_j, $j = 1, \dots, r$, be a collection of continuous linear functionals on $L_1(\omega, \mathbb{R}^m)$. Since the restriction of such a function to a space of the scales (4.4) is continuous, the operator $B_r = B + \sum_{j=1}^r j_j(\cdot)\gamma_j$ can be regarded as a continuous finite-dimensional perturbation of B. Therefore, the operator B_r is compact on every space of the scales (4.4).

We consider the problem

$$u + B_r u = 0. \tag{4.8}$$

Let $u \in \widehat{L}_q(\omega, \mathbb{R}^m)$ be a solution to (4.8) with $q > 1$. Since $\gamma_j \in \widehat{C}^{k,\varepsilon}(\overline{\omega}, \mathbb{R}^m)$ and the operator B raises the smoothness by one, we can, using the theorems on embedding of $W_q^1(\omega)$ into $L_p(\omega)$ and arguing as in Lemma 2.1, prove that u belongs to the space $W_p^1(\omega, \mathbb{R}^m)$ for any p. For a sufficiently large p the space $W_p^1(\omega)$ is embedded into $C^{0,\mu}(\overline{\omega})$ with $\mu < 1$. Taking into account the increase of the smoothness by the operator B, but in the Hölder scale, we conclude that $u \in \widehat{C}^{k,\varepsilon}(\overline{\omega}, \mathbb{R}^m)$. Thus, any solution to the problem (4.8) of class $\widehat{L}_q(\omega, \mathbb{R}^m)$, $q > 1$, belongs to the space $\widehat{C}^{k,\varepsilon}(\overline{\omega}, \mathbb{R}^m)$. Therefore, the set of solutions to (4.8) is independent of the choice of the space from the scales (4.4).

We assume that we have found γ_j and l_j such that (see below on how this can be done) the problem (4.8) has only the zero solution in the space $H = \widehat{L}_2(\omega, \mathbb{R}^m)$. For these γ_j and l_j the problem (4.8) has only the zero solution in every space of the scale (4.4).

We consider the nonhomogeneous problem (4.8)

$$v + B_r v = w, \tag{4.9}$$

where w is the solution to (4.6). The set of solutions to (4.9) is independent of the choice of the space in the scales (4.4). The arguments leading to this conclusion are that every solution to the problem (4.9) of class $\widehat{L}_q(\omega, \mathbb{R}^m)$, $q > 1$, belongs to the space $\widehat{C}^{k,\varepsilon}(\overline{\omega}, \mathbb{R}^m)$. On every space of the scales (4.4), the operator B_r is compact and the problem (4.8) has only the zero solution. Therefore, the problem (4.9) is uniquely solvable and its solution (common for all the spaces) satisfies the estimates

$$|v|_{s-1,\varepsilon} \le C_s |w|_{s-1,\varepsilon}, \quad s = 1, \dots, k, \tag{4.10}$$

$$\|v\|_q \le C_q \|w\|_q, \quad 1 < q < \infty. \tag{4.11}$$

From (4.9) and the properties of B_r we immediately obtain

$$|v|_{s,\varepsilon} \le |B_r v|_{s,\varepsilon} + |w|_{s,\varepsilon} \le C_s |v|_{s-1,\varepsilon} + |w|_{s,\varepsilon}, \tag{4.12}$$

$$\|v\|_{1,q} \le \|B_r v\|_{1,q} + \|w\|_{1,q} \le C_q \|v\|_q + \|w\|_{1,q}, \tag{4.13}$$

where s and q are the same as above. From (4.6) we find that

$$|w|_{s,\varepsilon} \le C_s |g|_{s-1,\varepsilon}, \qquad \|w\|_{1,q} \le C_q \|g\|_q. \tag{4.14}$$

Combining (4.12), (4.10), and (4.14), we arrive at the first estimate in (4.3). The second estimate follows from (4.13), (4.11), and (4.14). We note that the solution depends linearly on the right-hand side g.

We now construct γ_j and l_j. If the problem $u + Bu = 0$ has only the zero solution in a Hilbert space H, it is possible to dispense with the empty collections of γ_j and l_j. Assume that the problem $u+Bu = 0$ has a nonzero solution in H. Let $N = \ker(I+B)$ and $R = \operatorname{Im}(I+B)$, where I denotes the unit operator on H. Since B is compact in H, we see that the subspace R is closed and $\dim N = \operatorname{codim} R = r$, $1 \leq r < \infty$.

We begin with the following remark. Let u and γ run over the set of all smooth functions with compact support and over the set of all smooth solenoidal vector-valued functions with compact support respectively. We prove that the set $\{u + Bu + \gamma\}$ is dense in H. It suffices to show that any $\beta \in H$ such that

$$\langle u + Bu + \gamma, \beta\rangle = 0 \tag{4.15}$$

vanishes for u and γ as above; here $\langle\cdot,\cdot\rangle = \int_\omega (\cdot,\cdot)_m\, dx$ denotes the inner product in H. In (4.15), we put $u = 0$ and $\gamma = \gamma_1 - \langle\sigma,\gamma_1\rangle\langle\sigma,\gamma\rangle^{-1}\gamma_0$, where $\gamma_1(x)$ is a smooth solenoidal vector-valued function with compact support and $\gamma_0(x)$ is a fixed smooth solenoidal vector-valued function with compact support such that $\langle\sigma,\gamma_0\rangle \neq 0$ (cf. Lemma 3.3). Then $\gamma \in H$, and (4.15) implies

$$0 = \big\langle \gamma_1 - \langle\sigma,\gamma_1\rangle\langle\sigma,\gamma_0\rangle^{-1}\gamma_0, \beta\big\rangle = \big\langle \gamma_1, \beta - \langle\gamma_0,\beta\rangle\langle\sigma,\gamma_0\rangle^{-1}\sigma\big\rangle.$$

Using Theorem 3.2, we can obtain the representation

$$\beta = \langle\beta,\gamma_0\rangle\langle\sigma,\gamma_0\rangle^{-1}\sigma + \nabla p,$$

where $p \in W_2^1(\omega)$. Since $\beta \in H$, we can take the inner product of the last equality and σ in $\mathbb{R}^m$, integrate the result over ω, and arrive at the relation $\langle\beta,\gamma_0\rangle = 0$, which yields $\beta = \nabla p$. Substituting β into (4.15) with $\gamma = 0$ and arbitrary u, we find that

$$0 = \langle u + Bu, \nabla p\rangle = -\int\limits_\omega (\operatorname{div} u + (\sigma, u)_m)\, p\, dx \int\limits_\omega (\nabla p - \sigma p, u)_m\, dx.$$

Therefore, $p(x)$ satisfies the equation $-\nabla p + \sigma p = \alpha\sigma$, $\alpha \in \mathbb{R}^1$, which can be written as

$$-\nabla(p - \alpha) + \sigma(p - \alpha) = 0 \tag{4.16}$$

for any α. The left-hand side of (4.16) coincides with $L^*(p-\alpha)$, where L is defined in (4.1). By Lemma 2.1, any solution to (4.16) of class $W_2^1(\omega)$ belongs to $C^1(\overline{\omega})$ and, by Theorem 2.1 ($\sigma \not\equiv 0$), it vanishes. Thus, $p \equiv \alpha$ and $\beta = \nabla p \equiv 0$, which proves the density.

Let $f_1, \dots, f_r$ and $e_1, e_2, \dots$ be orthonormal bases for $H \ominus R$ and R respectively. Their union also forms an orthonormal basis for H. From the above assertion about density it follows that for every $\delta > 0$, there exist $\widehat{f_j} = u_j + Bu_j + \gamma_j$, $j = 1, \dots, r$, such that $\|f_j - \widehat{f_j}\|_H < \delta$. We define a linear operator $F : H \to H$ acting on the basis $f_1, \dots, f_r, e_1, e_2, \dots$ by setting $Ff_j = \widehat{f}_j$ for $j = 1, \dots, r$ and $Fe_j = e_j$ for $j = 1, 2, \dots$. If

$$f = \sum_{j=1}^{r} c_j f_j + \sum_{j=1}^{\infty} \alpha_j e_j$$

is the decomposition of $f \in H$, then

$$\|Ff - f\|_H = \left\| \sum_{j=1}^{r} c_j \left(\widehat{f}_j - f_j \right) \right\|_H \le \delta r^{1/2} \left(\sum_{j=1}^{r} |c_j|^2 \right)^{1/2} \le \delta r^{1/2} \|f\|_H$$

and $\|F - I\| < 1$ for $\delta r^{1/2} < 1$. Thus, $F = I + (F - I)$, being a small perturbation of the identity operator, is an isomorphism of H. In particular, this means that any element $\varphi \in H$ is located in the range of F, which is equal to H, i.e., for some $f \in H$

$$\varphi = Ff = \sum_{j=1}^{r} c_j \widehat{f}_j + \widehat{\varphi}' = \sum_{j=1}^{r} c_j \gamma_j + \varphi'$$

with some $\widehat{\varphi}' \in R$ and $\varphi' \in R$. Thus, any $\varphi \in H$ can be represented as follows:

$$\varphi = \sum_{j=1}^{r} c_j \gamma_j + \varphi', \qquad \varphi' \in R. \tag{4.17}$$

Let P be the orthogonal projection on $H \ominus R$ in H. Since $\varphi \in H$ is arbitrary, $P\varphi$ may take any value in the r-dimensional space $H \ominus R$. From (4.17) we find that $P\varphi = P(\varphi - \varphi') = \sum_{j=1}^{r} c_j P\gamma_j$. Since j varies precisely between 1 and r, the functions $P\gamma_j$ are linearly independent, and so the γ_j are linearly independent and do not belong to R. It is easy to prove the uniqueness of the representation (4.17). It suffices to prove that

$$\sum_{j=1}^{r} c_j \gamma_j + \varphi' = 0 \tag{4.18}$$

implies $c_j = 0$, $j = 1, \dots, r$, $\varphi' = 0$. Assume (4.18) holds. Then $\sum_{j=1}^{r} c_j P\gamma_j = 0$. By linear independence of $P\gamma_j$, every c_j is zero. Hence $\varphi' = 0$. Thus, (4.17) is a direct sum decomposition:

$$H = R \dotplus R', \tag{4.19}$$

where R' is the r-dimensional space spanned on γ_j, $j = 1, \dots, r$.

Let $v_1, \dots, v_r$ be a basis for the space N. We define functionals $\widehat{l}_j$, $j = 1, \dots, r$, on N (it suffices to define them only on the basis) as follows: $\widehat{l}_j(v_i) = \delta_{ij}$. Since $N \subset \widehat{L}_1(\omega, \mathbb{R}^m)$, the Hahn–Banach theorem shows that every functional $\widehat{l}_j$ admits an extension to a continuous functional l_j defined on $\widehat{L}_1(\omega, \mathbb{R}^m)$. Arguing as above, we define B_r using γ_j and l_j. Let $u \in H$ be a solution to (4.8). Since (4.19) is a direct sum decomposition, u satisfies the equalities

$$u + Bu = 0, \qquad \sum_{j=1}^{r} l_j(u) \gamma_j = 0. \tag{4.20}$$

The first equality in (4.20) means that $u \in N$, i.e., $u = \sum_{j=1}^{r} c_j v_j$, $c_j \in \mathbb{R}^1$. Since the γ_j are linearly independent, the second equality in (4.20) yields

$$0 = l_j(u) = l_j\left(\sum_{s=1}^{r} c_s v_s\right) = c_j, \qquad j = 1, \dots, r,$$

whence $u = 0$. Thus, the l_j and the γ_j constitute the required system of functions and vectors. □

Lemma 4.1 provides the background to the study of the following more general problem for the operator (4.1) under the condition that σ is not identically zero:

$$(4.21) \qquad \begin{aligned} Lu &= g, & u\big|_{\partial\omega} &= \varphi, \\ g &\in C^{k-1,\varepsilon}(\overline{\omega}), & \varphi &\in C^{k,\varepsilon}(\partial\omega, \mathbb{R}^m). \end{aligned}$$

THEOREM 4.1. *Let $\omega \subset \mathbb{R}^m$ be a bounded domain with boundary $\partial\omega \in C^{k+1,\varepsilon}$, $k \geq 1$, $\varepsilon \in (0,1]$. For any g and φ the problem (4.21) has a solution $u \in C^{k,\varepsilon}(\overline{\omega}, \mathbb{R}^m)$ that depends linearly on g, φ and satisfies the estimates*

$$(4.22) \qquad \begin{aligned} |u|_{s,\varepsilon} &\leq C_s\left(|g|_{s-1,\varepsilon} + |\varphi|_{s,\varepsilon}\right), \quad s = 1, \dots, k, \\ \|u\|_{1,q} &\leq C_q\left(\|g\|_q + \|\varphi\|_{1-1/q,q}\right), \quad q \in (1, \infty). \end{aligned}$$

PROOF. We seek a solution to (4.21) in the form $u = v + w + c\sigma$, where v, w are unknown functions and c is a constant. The problem (4.21) can be rewritten as follows:

$$(4.23) \qquad \begin{aligned} Lv + (\sigma, w)_m &= g - c|\sigma|^2 - \operatorname{div} w, \\ v\big|_{\partial\omega} &= (\varphi - c\sigma - w)\big|_{\partial\omega}, \end{aligned}$$

For w we take a solution to the problem

$$(4.24) \qquad \begin{aligned} \operatorname{div} w &= g - c|\sigma|^2, & w &\in C^{k,\varepsilon}(\overline{\omega}, \mathbb{R}^m), \\ \int_\omega (\sigma, w)_m \, dx &= 0, & w\big|_{\partial\omega} &= (\varphi - c\sigma)\big|_{\partial\omega}. \end{aligned}$$

The criterion for solvability of (4.24) is given by the relation

$$\int_\omega (g - c|\sigma|^2) \, dx = \int_{\partial\omega} \varphi_\nu \, dx,$$

from which the value of c can be determined:

$$(4.25) \qquad c = \|\sigma\|_2^{-2}\left(\int_\omega g \, dx - \int_{\partial\omega} \varphi_\nu \, dS\right).$$

It is obvious that such c depends linearly on g, φ and satisfies the estimates

$$(4.26) \qquad \begin{aligned} |c| &\leq C_s(|g|_{s-1,\varepsilon} + |\varphi|_{s,\varepsilon}), \\ |c| &\leq C_q(\|g\|_q + \|\varphi\|_{1-1/q,q}), \end{aligned}$$

where $s = 1, \dots, k$ and $q \in (1, \infty)$. If (4.25) holds, we apply Lemma 3.3 to the problem (4.24). By Lemma 3.3 and the estimates (4.26), the problem (4.24) has a solution depending linearly on g, φ and, for $s=1,\dots,k$, $q\in(1,\infty)$, satisfying the estimates

$$|w|_{s,\varepsilon}\le C_s\left(|g|_{s-1,\varepsilon}+|\varphi|_{s,\varepsilon}\right), \tag{4.27}$$
$$\|w\|_{1,q}\le C_q\left(\|g\|_q+\|\varphi\|_{1-1/q,q}\right).$$

To find v it remains to use Lemma 4.1 in order to solve the problem $Lv=-(\sigma,w)_m$, $v\big|_{\partial\omega}=0$. From Lemma 4.1 and the estimates (4.27) we obtain the conclusion of the theorem. □

§5. Boundary-value problems for the operator L

Let $\omega\subset\mathbb{R}^m$ be a bounded domain with boundary $\partial\omega\in C^{k+1,\varepsilon}$, where k is a natural number and $\varepsilon\in(0,1]$. For given $f\in C^{k-1,\varepsilon}(\overline{\omega})$ and $\varphi\in C^{k,\varepsilon}(\partial\omega,\mathbb{R}^n)$ we formulate the boundary-value problem

$$Lu=f,\quad u\big|_{\partial\omega}=\varphi,\quad u\in C^{k,\varepsilon}(\overline{\omega},\mathbb{R}^n). \tag{5.1}$$

THEOREM 5.1. (a) *If the problem* (2.2) *has only the zero solution, then the problem* (5.1) *is solvable for any* f *and* φ.

(b) *If the problem* (2.2) *has a nonzero solution, then the problem* (5.1) *is solvable for those and only those* f *and* φ *for which*

$$\int_\omega p^*f\,dx=\int_{\partial\omega}p^*(A\nu,\varphi)_n\,dS. \tag{5.2}$$

(c) *In both cases* (a) *and* (b), *we can find a solution that depends linearly on* f, φ *and satisfies the estimates*

$$|u|_{s,\varphi}\le\left(|f|_{s-1,\varphi}+|\varphi|_{s,\varphi}\right),\qquad \|u\|_{1,q}\le C_q\left(\|f\|_q+\|\varphi\|_{1-1/q,q}\right)$$

for $s=1,\dots,k$ *and* $q\in(1,\infty)$.

PROOF. We represent functions $u\in C^{k,\varepsilon}(\overline{\omega},\mathbb{R}^n)$ and $\varphi\in C^{k,\varepsilon}(\partial\omega,\mathbb{R}^n)$ in the form

(5.3)
$$u=e^{-\psi}AG^{-1}v+w,\quad v\in C^{k,\varepsilon}(\overline{\omega},\mathbb{R}^m),\quad w\in C^{k,\varepsilon}(\overline{\omega},\mathbb{R}^n),\quad A^*w\equiv 0,$$
$$\varphi=e^{-\psi}AG^{-1}\varphi_1+\varphi_2,\quad \varphi_1\in C^{k,\varepsilon}(\partial\omega,\mathbb{R}^m),\quad \varphi_2\in C^{k,\varepsilon}(\partial\omega,\mathbb{R}^n),\quad A^*\varphi_2\equiv 0.$$

Such a representation exists and is unique:

$$v=e^{\psi}A^*u,\quad w=u-AG^{-1}A^*u, \tag{5.4}$$
$$\varphi_1=e^{\psi}A^*\varphi,\quad \varphi_2=\varphi-AG^{-1}A^*\varphi.$$

Recall that the functions ψ, b, and σ were defined in (1.5). An easy computation yield the equality

$$Lu=e^{-\psi}\left[\operatorname{div}v+(\sigma,v)_m\right]+(b,w)_n.$$

Taking into account the uniqueness of the representations in (5.3) and the expression for Lu, we can equivalently rewrite the problem (5.1) as follows:

$$\operatorname{div}v+(\sigma,v)_m=e^{\psi}\left[f-(b,w)_n\right],\quad A^*w=0, \tag{5.5}$$
$$v\big|_{\partial\omega}=\varphi_1,\quad w\big|_{\partial\omega}=\varphi_2.$$

The functions v and w are unknown in this problem.

We begin by constructing w. The function w must be chosen so as to satisfy the conditions

$$A^*w \equiv 0, \quad w\big|_{\partial\omega} = \varphi_2, \quad w \in C^{k,\varepsilon}(\overline{\omega}, \mathbb{R}^n); \tag{5.6}$$

moreover, the problem with respect to v must be solvable. The solvability condition depends on σ. In the case $\sigma \not\equiv 0$, no additional conditions are required (cf. Theorem 4.1), but in the case $\sigma \equiv 0$ we must impose the following additional condition (cf. Theorem 3.1):

$$\int_\omega e^\psi \left[f - (b, w)_n\right] dx = \int_{\partial\omega} \varphi_{1\nu}\, dS = \int_{\partial\omega} e^\psi (A\nu, \varphi)_n\, dS. \tag{5.7}$$

Let $\Pi : C^{k,\varepsilon}(\partial\omega, \mathbb{R}^n) \to C^{k,\varepsilon}(\overline{\omega}, \mathbb{R}^n)$ be the extension operator (cf. the Appendix) such that

$$|\Pi\varphi|_{s,\varepsilon} \le C_s |\varphi|_{s,\varepsilon}, \quad \|\Pi\varphi\|_{1,q} \le C_q \|\varphi\|_{1-1/q,q} \tag{5.8}$$

for $s = 1, \dots, k$, $q \in (1, \infty)$, and $\varphi \in C^{k,\varepsilon}(\partial\omega, \mathbb{R}^n)$. For $b \not\equiv 0$ we introduce a function $\zeta \in C_0^\infty(\omega)$ such that

$$\zeta(x) \ge 0, \quad \operatorname{supp}\zeta(x) \cap \operatorname{supp} b(x) \ne \varnothing.$$

We modify the extension operator Π by the following formulas:

$$\Pi'\varphi = \Pi\varphi \quad \text{if} \quad b \equiv 0, \qquad \Pi'\varphi = \Pi\varphi - \alpha\zeta(x)e^\psi b \quad \text{if} \quad b \not\equiv 0,$$

where

$$\alpha = \left(\int_\omega (\Pi\varphi, b)_n e^\psi\, dx\right)\left(\int_\omega \zeta(x)e^{2\psi}|b|^2\, dx\right)^{-1}.$$

The estimates (5.8) remain valid for Π'; moreover, the integral of $e^\psi(\Pi'\varphi, b)_n$ over ω vanishes. We define

$$w(x) = \Pi'\varphi_2 - AG^{-1}A^*\Pi'\varphi_2 + c\zeta b, \quad c \in \mathbb{R}^1. \tag{5.9}$$

Such a function $w(x)$ satisfies (5.6). We define c by the relation

$$c = \left[\int_\omega e^\psi f\, dx - \int_{\partial\omega} e^\psi (A\nu, \varphi)_n\, dS\right]\left[\int_\omega e^\psi \zeta |b|^2\, dx\right]^{-1}$$

if $b \not\equiv 0$, and put $c = 0$ if $b \equiv 0$. Then the condition (5.7) is fulfilled for $b \not\equiv 0$. By the estimates for Π' and the representation of φ_2 in (5.4), the function $w(x)$ depends linearly on f, φ and satisfies the estimates

$$\begin{aligned} |w(x)|_{s,\varepsilon} &\le C_s \left(\|f\|_1 + |\varphi|_{s,\varepsilon}\right), \\ \|w\|_{1,q} &\le C_q \left(\|f\|_1 + \|\varphi\|_{1-1/q,q}\right). \end{aligned} \tag{5.10}$$

for $s = 1, \dots, k$ and $q \in (1, \infty)$.

We now define v. In the case (a), $|\sigma(x)| + |b(x)| \not\equiv 0$ (cf. Theorem 2.1). Hence the solvability condition for the problem (5.5) with respect to v is fulfilled. The solution to this problem from Theorem 3.1 ($\sigma \equiv 0$) or Theorem 4.1 ($\sigma \not\equiv 0$) depends linearly on φ_1, f, w and satisfies the estimates

$$
\begin{aligned}
|v|_{s,\varepsilon} &\le C_s \left(|f|_{s-1,\varepsilon} + |w|_{s-1,\varepsilon} + |\varphi_1|_{s,\varepsilon}\right), \\
\|v\|_{1,q} &\le C_q \left(\|f\|_q + \|w\|_q + \|\varphi_1\|_{1-1/q,q}\right),
\end{aligned}
\tag{5.11}
$$

where $s = 1, \ldots, k$ and $q \in (1, \infty)$. From the linear dependence of w on f, φ, the estimates (5.10), (5.11), and the representation of φ_1 in (5.4) we conclude that v linearly depends on f, φ and satisfies the following estimates:

$$
\begin{aligned}
|v|_{s,\varepsilon} &\le C_s \left(|f|_{s-1,\varepsilon} + |\varphi|_{s,\varepsilon}\right), \\
\|v\|_{1,q} &\le C_q \left(\|f\|_q + \|\varphi\|_{1-1/q,q}\right),
\end{aligned}
\tag{5.12}
$$

where s and q are as above. Using the representation of u in (5.3), we find a solution possessing the properties required in (c).

In the case (b), $\sigma \equiv 0$ and $b \equiv 0$ (cf. Theorem 2.1). Then the relation (5.7), which holds by assumption, coincides with (5.2). Repeating the above arguments (but with Theorem 3.1, since now $\sigma \equiv 0$), we construct a solution possessing the properties required in (c). This proves the sufficiency of the condition (5.2). To prove the necessity, we multiply the equality (5.1) by p^* and integrate the result by parts. □

To consider boundary conditions more general than in (5.1), we need some notions. Let $\Gamma \subset \partial\omega$ be an open set of $\partial\omega$ and let its boundary $\partial\Gamma$ consist of a finite number of components of class $C^{k,\varepsilon}$ (cf. the Appendix). The cases $\Gamma = \varnothing$ and $\Gamma = \partial\omega$ are possible. Finally, let $S = \partial\omega \backslash \overline{\Gamma}$. On $\partial\omega$, we fix a field $\gamma \in C^{k,\varepsilon}(\partial\omega, \mathbb{R}^n)$ of unit length and for given $f \in C^{k-1,\varepsilon}(\overline{\omega})$ and $\varphi \in C^{k,\varepsilon}(\partial\omega, \mathbb{R}^n)$ consider the following problem: to find a function $u \in C^{k,\varepsilon}(\overline{\omega}, \mathbb{R}^n)$ such that

$$
Lu = f, \quad u\big|_\Gamma = \varphi\big|_\Gamma, \quad (u, \gamma)_n = (\varphi, \gamma)_n\big|_S.
\tag{5.13}
$$

A problem with no boundary condition imposed on Γ and (or) S will be formally mentioned as (5.13) with $\Gamma = *$ and (or) $S = *$. Under the same assumptions on ω we prove the following theorem.

THEOREM 5.2. (a) *If the problem* (2.2) *has only the zero solution, then for any S and Γ the problem* (5.13) *has a solution that depends linearly on f, φ and satisfies the estimates*

$$
\begin{aligned}
|u|_{s,\varepsilon} &\le C_s \left[|f|_{s-1,\varepsilon} + |\varphi|_{s,\varepsilon}(\Gamma) + |(\varphi, \gamma)_n|_{s,\varepsilon}(S)\right], \\
\|u\|_{1,q} &\le C_q \left[\|f\|_q + \|\varphi\|_{1-1/q,q}(\Gamma) + \|(\varphi, \gamma)_n\|_{1-1/q,q}(S)\right]
\end{aligned}
\tag{5.14}
$$

for $s = 1, \ldots, k$ and $q \in (1, \infty)$. If S and (or) Γ are $*$, then the corresponding norms on the right-hand side of (5.14) *are absent.*

(b) *If the problem* (2.2) *has a nonzero solution, but one of the following conditions holds*:

(b$_1$) $\Gamma \ne \partial\omega$, $S = *$,

(b$_2$) $\Gamma = *$, $S \ne \partial\omega$,

(b$_3$) $* \ne \Gamma \ne \partial\omega$, $S \ne *$, $A(x_0)_\nu(x_0) \ne \beta\gamma(x_0)$ *for some $x_0 \in S$ and all $\beta \in \mathbb{R}^1$, then the claim* (a) *is true.*

(c) *If the problem* (2.2) *has a nonzero solution and none of the conditions* (b_1)–(b_3) *is satisfied, i.e., one of the following conditions holds*:

(c_1) $* \neq \Gamma = \partial\omega$,

(c_2) $* \neq \Gamma \neq \partial\omega$, $A(x)\nu(x) \equiv \beta(x)\gamma(x)$ *for* $x \in S \neq *$ *with some scalar function* $\beta(x)$,

then the equality (5.2) *is a necessary and sufficient condition for the solvability of* (5.13). *If* (5.2) *holds, the problem* (5.13) *has a solution possessing the same properties as indicated in* (a).

PROOF. Any solution to the problem (5.1) satisfies the problem (5.13). However, the solvability condition for (5.1) is stronger than the condition stated for (5.13). This can be explained by the fact that only the component of φ along γ appears in the boundary condition on S in (5.13), but not the entire function φ. Therefore, we can try to modify the function φ on S in such a way that the component of φ along γ remains unchanged but the problem (5.1) becomes solvable. The principal idea is to find a suitable modification of φ.

We consider the case in which $\varnothing \neq \Gamma \neq \partial\omega$ and neither Γ nor S takes the formal value $*$. Let $\Pi : C^{k,\varepsilon}(\Gamma, \mathbb{R}^n) \to C^{k,\varepsilon}(\partial\omega, \mathbb{R}^n)$ be an extension operator such that

$$|\Pi\varphi|_{s,\varepsilon}(\partial\omega) \leq C_s|\varphi|_{s,\varepsilon}(\Gamma), \quad \|\Pi\varphi\|_{1-1/q,q}(\partial\omega) \leq C_q\|\varphi\|_{1-1/q,q}(\Gamma)$$

(cf. the Appendix), where $s = 1, \ldots, k$ and $q \in (1, \infty)$. Starting from the function $\varphi \in C^{k,\varepsilon}(\partial\omega, \mathbb{R}^n)$, we define $\varphi_1 \in C^{k,\varepsilon}(\partial\omega, \mathbb{R}^n)$ by the equality

$$\varphi_1 = \Pi\{[\varphi - \gamma(\varphi, \gamma)_n]\big|_\Gamma\}$$

and introduce the operator $\widehat{\Pi} : C^{k,\varepsilon}(\partial\omega, \mathbb{R}^n) \to C^{k,\varepsilon}(\partial\omega, \mathbb{R}^n)$ as follows:

$$\widehat{\Pi}\varphi = \varphi_1 - \gamma(\varphi_1, \gamma)_n + \gamma(\varphi, \gamma)_n.$$

The operator $\widehat{\Pi}$ satisfies the conditions

$$\begin{gathered} \widehat{\Pi}\varphi\big|_\Gamma = \varphi\big|_\Gamma, \qquad (\widehat{\Pi}\varphi, \gamma)_n\big|_S = (\varphi, \gamma)_n\big|_S, \\ |\widehat{\Pi}\varphi|_{s,\varepsilon} \leq C_s\left[|\varphi|_{s,\varepsilon}(\Gamma) + |(\varphi, \gamma)_n|_{s,\varepsilon}(S)\right], \quad s = 1, \ldots, k, \\ \|\widehat{\Pi}\varphi\|_{1-1/q,q} \leq C_q\left[\|\varphi\|_{1-1/q,q}(\Gamma) + \|(\varphi, \gamma)_n\|_{1-1/q,q}(S)\right], \quad q \in (1, \infty). \end{gathered} \tag{5.15}$$

Let the assumptions of (a) be satisfied. Then, for a solution to problem (5.13) with the properties indicated in (a), we can take the solution u to the problem

$$Lu = f, \qquad u\big|_{\partial\omega} = \widehat{\Pi}\varphi \tag{5.16}$$

from Theorem 5.1. If (b_3) holds, then there exists a function $\varphi_0 \in C^{k,\varepsilon}(\partial\omega, \mathbb{R}^n)$ whose support lies in a small neighborhood of x_0 and

$$(\varphi_0(x), \gamma(x))_n \equiv 0, \qquad \int\limits_{\partial\omega} p^*(A\nu, \varphi_0)_n \, dS = 1. \tag{5.17}$$

We consider the boundary-value problem (5.16) with $\widehat{\Pi}\varphi$ replaced by $\widehat{\Pi}\varphi + c\varphi_0$. To satisfy the solvability condition (5.2), we define

$$c = c[f, \varphi] = \int\limits_{\omega} p^* f \, dx - \int\limits_{\partial\omega} p^*(A\nu, \widehat{\Pi}\varphi)_n \, dS. \tag{5.18}$$

Then the solution to the problem (5.16) constructed in Theorem 5.1 satisfies all the required conditions.

If (c_2) is satisfied, then the necessity of (5.2) can be obtained by integrating by parts the equality $p^* Lu = p^* f$ and taking into account the boundary conditions in (5.13), the equality in (c_2), and the relation $L^* p^* \equiv 0$. If (5.2) holds for the pair f, φ, then it is also valid for the pair f, $\widehat{\Pi}\varphi$. Hence the solution to the problem (5.16) from Theorem 5.1(b) satisfies the required conditions.

Consider the case in which $\varnothing \neq \Gamma \neq \partial\omega$, but either Γ or S takes the value $*$. We assume that $S = *$ and $\Gamma \neq *$. Let Π be the extension operator. If the assumptions of the claim (a) are satisfied, then for a solution to (5.13) we can take the solution to (5.16) with Π instead of $\widehat{\Pi}$. If (b_1) is satisfied, then the boundary condition in (5.16) must be replaced by the following condition:

$$u\big|_{\partial\omega} = \Pi\varphi + c\varphi_0, \quad \varphi_0 \in C^{k,\varepsilon}(\partial\omega, \mathbb{R}^n), \quad \operatorname{supp}\varphi_0 \subset S;$$

moreover, φ_0 satisfies the second equality in (5.17) and c is defined by (5.18) with $\widehat{\Pi}$ replaced by Π. For the problem (5.16) modified in such a way, a solution exists and possesses the required properties.

We now assume that $S \neq *$ and $\Gamma = *$. In this case, to prove (a) and (b_2) one can repeat the above arguments interchanging Γ and S.

We now study the case $\Gamma = \varnothing$ and $S \neq *$. If the assumptions of (a) are satisfied, for a solution to (5.13) we can take the solution to the problem

$$Lu = f, \qquad u\big|_{\partial\omega} = \gamma(\varphi, \gamma)_n$$

from Theorem 5.1. If (b_3) is satisfied, then (5.17) is satisfied by the solution to the problem

$$Lu = f, \qquad u\big|_{\partial\omega} = \gamma(\varphi, \gamma)_n + c\varphi_0,$$

where $\varphi_0 \in C^{k,\varepsilon}(\partial\omega, \mathbb{R}^n)$ and the support of φ belongs to a neighborhood of x_0; moreover,

$$c = \int\limits_{\omega} p^* f \, dx - \int\limits_{\partial\omega} p^*(A\nu, \gamma)_n (\varphi, \gamma)_n \, dx$$

has all the required properties.

The case $\Gamma = \varnothing$ and $S = *$ corresponds to the problem for the equation $Lu = f$ without boundary conditions. As in (a) and (b_1), for the required solution to (5.13) we can take a solution to the problem $Lu = f$, $u\big|_{\partial\omega} = c\varphi_0$, where φ_0 is of class $C^{k,\varepsilon}(\partial\omega, \mathbb{R}^n)$ and satisfies the second equality in (5.17) while c is equal to the integral of $p^* f$ over ω.

In the case $* \neq \Gamma = \partial\omega$, Theorems 5.1 and 5.2 coincide. $\square$

§6. The kernel of the operator L

Let $\omega \subset \mathbb{R}^m$ be a bounded domain with boundary $\partial\omega \in C^{k,\varepsilon}$, $k \geq 1$, $\varepsilon \in (0,1]$. By $\widehat{C}^{s,\varepsilon}(\overline{\omega}, \mathbb{R}^n)$, $s = 1, \ldots, k$, we denote the completion of the space $C^{k,\varepsilon}(\overline{\omega}, \mathbb{R}^n)$ in the norm $|\cdot|_{s,\varepsilon}$. We note that $\widehat{C}^{k,\varepsilon}(\overline{\omega}, \mathbb{R}^n) = C^{k,\varepsilon}(\overline{\omega}, \mathbb{R}^n)$ and $C^{k,\varepsilon}(\overline{\omega}, \mathbb{R}^n)$ is dense in $W^1_q(\overline{\omega}, \mathbb{R}^n)$. Introduce the spaces $J^{s,\varepsilon}(\Gamma, S)$ and $J_{1,q}(\Gamma, S)$ as follows:

$$
\begin{aligned}
J^{s,\varepsilon}(\Gamma, S) &= \left\{u \in C^{s,\varepsilon}(\overline{\omega}, \mathbb{R}^n) : Lu = 0,\ u\big|_\Gamma = 0,\ (u,\gamma)_n\big|_S = 0\right\}, \\
J_{1,q}(\Gamma, S) &= \left\{u \in W^1_q(\omega, \mathbb{R}^n) : Lu = 0,\ u\big|_\Gamma = 0,\ (u,\gamma)_n\big|_S = 0\right\},
\end{aligned}
\tag{6.1}
$$

where Γ, S, and γ are the same as in §5. We begin the study the properties of the spaces (6.1) with the proof of their nontriviality. The proof is based on a number of preliminary constructions.

LEMMA 6.1. *For any $\zeta_0 \in \mathbb{R}^n$ and $x_0 \in \omega$ there exists a family of smooth n-dimensional vector-valued functions $v_\tau(x)$ with compact support such that*

$$
\begin{aligned}
&v_\tau(x_0) = \zeta_0, \quad \operatorname{supp} v_\tau(x) \text{ contracts to a point } x_0 \text{ as } \tau \to 0, \\
&\qquad \lim_{\tau\to 0} \|Lv_\tau\|_q = 0 \quad \text{for some } q > m.
\end{aligned}
\tag{6.2}
$$

PROOF. We represent a vector $x \in \mathbb{R}^m$ in the form

$$x = x_0 + t\lambda + y, \quad \lambda \in \mathbb{R}^m, \quad |\lambda| = 1, \quad t \in \mathbb{R}^1, \quad y \in \mathbb{R}^m, \quad (y,\lambda)_m = 0,$$

where x_0 and λ are fixed. Since $t = (x - x_0, \lambda)_m$ and $y = x - x_0 - \lambda(x - x_0, \lambda)_m$, the functions t and y are infinitely differentiable and $t_{x_j} = \lambda_j$, $y_{ix_j} = \delta_{ij} - \lambda_i\lambda_j$. Consider the cylinder

$$Q_{l\tau} = \{x \in \mathbb{R}^m : |t| \leq l,\ |y| \leq \tau,\ 0 < l \leq \tau \leq 1\}.$$

Finally, let $\alpha \in C^\infty(\mathbb{R}^1_+)$, $\alpha(z) \equiv 1$ for $0 \leq z \leq 1/2$, $\alpha(z) \equiv 0$ for $z \geq 1$, $\zeta \in C^\infty(\mathbb{R}^m, \mathbb{R}^n)$, and $\zeta(x_0) = \zeta_0$. We define a two-parameter family $v_{l\tau}(x)$ of smooth n-dimensional vector-valued functions with compact support as follows:

$$v_{l\tau}(x) = \zeta(x)\alpha(|t|/l)\alpha(|y|/\tau), \quad 0 < l \leq \tau \leq 1.$$

Since the support of $v_{l\tau}(x)$ belongs to $Q_{l\tau}$, for sufficiently small τ the supports of functions of the family $v_{l\tau}$ lie in ω. Using the relations

$$
\begin{aligned}
\frac{\partial v^i_{l\tau}}{\partial x_j} &= \zeta^i_{x_j}\alpha\left(\frac{|t|}{l}\right)\alpha\left(\frac{|y|}{\tau}\right) + l^{-1}\zeta^i\lambda_j\dot{\alpha}\left(\frac{|t|}{l}\right)\alpha\left(\frac{|y|}{\tau}\right)\operatorname{sgn} t \\
&\quad + \tau^{-1}\zeta^i y_j\dot{\alpha}\left(\frac{|y|}{\tau}\right)\alpha\left(\frac{|t|}{l}\right)|y|^{-1},
\end{aligned}
$$

it is easy to obtain the formula

$$
\begin{aligned}
Lv_{l\tau} &= l^{-1}(A^*\zeta, \lambda)_m\dot{\alpha}\left(\frac{|t|}{l}\right)\alpha\left(\frac{|y|}{\tau}\right)\operatorname{sgn} t + \tau^{-1}(A^*\zeta, y)_m\dot{\alpha}\left(\frac{|y|}{\tau}\right)\alpha\left(\frac{|t|}{l}\right)|y|^{-1} \\
&\quad + \left(a^i + \frac{\partial A_{ij}}{\partial x_j}\right)\zeta^i\alpha\left(\frac{|t|}{l}\right)\alpha\left(\frac{|y|}{\tau}\right) + A_{ij}\zeta^i_{x_j}\alpha\left(\frac{|t|}{l}\right)\alpha\left(\frac{|y|}{\tau}\right),
\end{aligned}
$$

which implies the estimate

$$|Lv_{l\tau}(x)| \leq C\left[l^{-1}|(A^*(x)\zeta(x), \lambda)_m| + \tau^{-1} + 1\right], \qquad C \neq C(l,\tau)$$

for $x \in Q_{lq} \cap \omega$. We fix λ so that the equality $(A^*(x_0)\zeta_0, \lambda)_m = 0$ will be satisfied. Taking into account that $\zeta(x_0) = \zeta_0$, the diameter of $Q_{l\tau}$ can be estimated in terms of τ, and the function $(A^*(x)\zeta(x), \lambda)_m$ is continuously differentiable, we see that the inequality $|(A^*(x)\zeta(x), \lambda)| \leq C\tau$ holds for $x \in Q_{l\tau} \cap \omega$. Therefore,

$$|Lv_{l\tau}(x)| \leq C\left(\tau l^{-1} + \tau^{-1} + 1\right), \quad x \in \omega \cap Q_{l\tau}.$$

Since $Lv_{l\tau} = 0$ in the remaining part of ω, we get

$$\|Lv_{l\tau}\|_q \leq C \operatorname{meas}^{1/q} Q_{l\tau} \left(\tau l^{-1} + \tau^{-1} + 1\right).$$

We set $l = \tau^{\varkappa}$, $\varkappa \geq 1$. Since $\operatorname{meas} Q_{l\tau}$ is estimated from above in terms of $Cl\tau^{m-1}$, the right-hand side of the last inequality does not exceed

$$C\left[\tau^{\varkappa+m-1-q(\varkappa-1)} + \tau^{\varkappa+m-1-q} + \tau^{\varkappa+m-1}\right]^{1/q}$$

with some positive constant C. The quantity tends to zero as $\tau \to 0$ only if all the exponents at powers of τ are positive. Analyzing these exponents (from the point of view of the maximization of q), we obtain the optimal value $\varkappa = 2$. In this case, all the exponents are positive for $q < m + 1$. Thus, $v_\tau(x) = v_{l\tau}(x)$ for $l = \tau^2$ satisfies all the requirements of the lemma. □

In the proof of Lemma 6.1, the structure of the operator L is essentially used. If we replace $\|Lv_\tau\|_q$ with $\|v_\tau\|_{1,q}$ in (6.2), the lemma becomes false. In the sequel, we will use not only the conclusion of the lemma, but the construction of the functions $v_{l\tau}(x)$ and $v_\tau(x)$ as well.

LEMMA 6.2. (a) *For any $x_0 \in \partial\omega$ and $\zeta_0 \in \mathbb{R}^n$ there exists a family of smooth vector-valued functions $v_\tau(x)$, $0 < \tau \leq 1$, with compact support in $\mathbb{R}^m$ such that their restrictions to ω satisfy* (6.2).

(b) *For any $x_0 \in \partial\omega$ and $\zeta_0 \in \mathbb{R}^n$ such that $(\zeta_0, \gamma(x_0))_n = 0$ there exists a family of vector-valued functions $v_\tau(x)$, $0 < \tau \leq 1$, with compact support in $\mathbb{R}^m$ which satisfy* (6.2) *and the condition $(v_\tau(x), \gamma(x))_n = 0$ for $x \in \partial\omega$.*

PROOF. The family $v_\tau(x)$ constructed in Lemma 6.1 satisfies all the requirements of (a). To prove (b), we extend the vector field $\gamma(x)$ from $\partial\omega$ to a vector field of class $C^{k,\varepsilon}(\mathbb{R}^m, \mathbb{R}^n)$ (cf. the Appendix). Using the same notation $\gamma(x)$ for the extended vector field, we modify the vector-valued function $\zeta(x)$ in the formula for $v_{l\tau}(x)$ by replacing it by $\widehat{\zeta}(x) = \zeta(x) - \gamma(x)(\zeta(x), \gamma(x))_n$. Then $(\widehat{\zeta}(x), \gamma(x))_n = 0$ for $x \in \partial\omega$. The modified family $v_\tau(x)$ satisfies all the requirements. □

Let $\overline{x}$ be a finite collection of points $x_j \in \overline{\omega}$, $j = 1, \dots, l$, and let $\overline{\zeta}$ be a collection of vectors $\zeta_j \in \mathbb{R}^n$, $j = 1, \dots, l$. If $x_j \in \overline{\Gamma} \neq *$ or $\overline{S} \neq *$ for a given j, we agree that the consistency condition $\zeta_j = 0$ or $(\zeta_j, \gamma(x_j))_n = 0$ holds respectively.

LEMMA 6.3. *There exists a family of n-dimensional vector-valued functions $v_\tau \in C^{k,\varepsilon}(\mathbb{R}^m, \mathbb{R}^n)$ such that for a sufficiently small $\tau > 0$ the following conditions are satisfied: $v_\tau(x_j) = \zeta_j$, $j = 1, \dots, l$, $\operatorname{supp} v_\tau(x)$ contracts to $\overline{x}$ as $\tau \to 0$, $\lim_{\tau \to 0} \|Lv_\tau\|_q = 0$ for some $q > m$, $v_\tau\big|_\Gamma = 0$, and $(v_\gamma, \gamma)_n\big|_S = 0$.*

PROOF. Let $v_\tau^{(j)}(x)$, $0 < \tau \le 1$, $j = 1, \dots, l$, be the family of functions constructed in Lemma 6.1 or Lemma 6.2 for $x_0 = x_j$. For $x_j \in \overline{\Gamma} \ne *$ we set $v_\tau^{(j)} \equiv 0$. Since for all $j' \ne j''$ and sufficiently small τ the supports of $v_\tau^{(j')}(x)$ and $v_\tau^{(j'')}(x)$ are mutually disjoint, $v_\tau = \sum_{j=1}^l v_\tau^{(j)}(x)$ satisfy all the requirements. □

The main result of this section is presented by the following theorem concerning the solvability of the problem

$$(6.3)\qquad \begin{aligned} &Lu = 0, \quad u\big|_\Gamma = 0, \quad (u, \gamma)_n\big|_S = 0, \\ &u(x_j) = \zeta_j, \quad j = 1, \dots, l, \qquad u \in C^{k,\varepsilon}(\overline{\omega}, \mathbb{R}^n). \end{aligned}$$

THEOREM 6.1. *The problem* (6.3) *has a solution that depends linearly on vectors of the collection* $\overline{\zeta}$ *and satisfies the estimate* $|u|_{k\varepsilon} \le C \sum_{j=1}^l |\zeta_j|$.

PROOF. We introduce the subspaces $R_j \subset \mathbb{R}^n$, $j = 1, \dots, l$, as follows:
$R_j = \mathbb{R}^n$ for $x_j \in \omega$, $x_j \in \Gamma = *$, $x_j \in S = *$, $x_j \in \partial\omega$ for $\Gamma = *$ and $S = *$,
$R_j = \{0\}$ for $x_j \in \overline{\Gamma} \ne *$,
$R_j = \{\zeta \in \mathbb{R}^n : (\zeta, \gamma(x_j))_n = 0\}$ for $x_j \in S \ne *$, $x_j \in \overline{S}$ for $S \ne *$ and $\Gamma = *$.
Then ζ belongs to the linear space $R = \prod_{j=1}^l R_j$ of dimension r. Let $\overline{\zeta}_1, \dots, \overline{\zeta}_r$ be a basis for R. For each of these collections $\overline{\zeta}_i$, $i = 1, \dots, r$, from the family of functions $v_\tau^{(i)}$ constructed in Lemma 6.3 and supplied with the subscript i, we introduce the family of functions $u_\tau^{(i)}$ that are solutions of the problems

$$\begin{gathered} Lu_\tau^{(i)}(x) = Lv_\tau^{(i)}(x), \quad u_\tau^{(i)} \in C^{k,\varepsilon}(\overline{\omega}, \mathbb{R}^n), \\ u_\tau^{(i)}\big|_\Gamma = v_\tau^{(i)}\big|_\Gamma = 0, \quad (u_\tau^{(i)}, \gamma)_n\big|_S = (v_\tau^{(i)}, \gamma)_n\big|_S = 0 \end{gathered}$$

from Theorem 5.2. It is easy to check that for a given right-hand side and fixed boundary conditions, the consistency condition in Theorem 5.2 is satisfied. By the boundedness of the embedding operator of $W_q^1(\omega)$ into $C(\overline{\omega})$ for $q > m$, these solutions satisfy the estimates

$$|u_\tau^{(i)}(x_j)| \le |u_\tau^{(i)}|_0 \le C_1 \|u_\tau^{(i)}\|_{1,q} \le C_2 \|Lv_\tau^{(i)}\|_q \to 0 \quad \text{as} \quad \tau \to 0,$$

where $i = 1, \dots, r$, $j = 1, \dots, l$. Then $w_\tau^{(i)}(x) = v_\tau^{(i)}(x) - u_\tau^{(i)} \in J^{k,\varepsilon}(\Gamma, S)$ for all $i = 1, \dots, r$ and the collections $\overline{\zeta}_i' = (w_\tau^{(i)}(x_1), \dots, w_\tau^{(i)}(x_l))$, $i = 1, \dots, r$, being a small perturbation of the basis $v_\tau \in C^{k,\varepsilon}(\overline{\omega}, \mathbb{R}^n)$, $0 < \tau \le 1$, in a finite-dimensional space, form a basis for the space R for small τ. We fix τ small enough.

We turn to the problem (6.3). Let $\overline{\zeta} \in R$ denote the collection of vectors $\zeta_j \in \mathbb{R}^n$, $j = 1, \dots, l$, and let $\overline{\zeta} = \sum_{i=1}^r c_i \overline{\zeta}_i'$ be its decomposition with respect to the basis $\overline{\zeta}_i'$, $i = 1, \dots, r$. Then $|c_i| \le C \sum_{j=1}^l |\zeta_j|$, $i = 1, \dots, r$. Hence $u(x) = \sum_{i=1}^r c_i w_\tau^{(i)}(x)$ satisfies all the requirements of the theorem. □

Combining Theorem 5.2 with Theorem 6.1, we obtain the following theorem.

THEOREM 6.2. *Under the assumptions of Theorem* 5.2 *and the consistency conditions* $\zeta_j = \varphi(x_j)$ *for* $x_j \in \overline{\Gamma} \ne *$ *and* $(\zeta_j, \gamma(x_j))_n = (\varphi(x_j), \gamma(x_j))_n$ *for* $x_j \in \overline{S} \ne *$, *the problem*

$$(6.4)\qquad Lu = f, \quad u\big|_\Gamma = \varphi\big|_\Gamma, \quad (u, \gamma)_n\big|_S, \quad u(x_j) = \zeta_j, \quad j = 1, \dots, l,$$

has a solution $u \in C^{k,\varepsilon}(\overline{\omega}, \mathbb{R}^n)$ *that depends linearly on the triple* f, φ, $\overline{\zeta}$ *and satisfies the estimates*

$$|u|_{s,\varepsilon} \le C_s \Big[|f|_{s-1,\varepsilon} + |\varphi|_{s,\varepsilon}(\Gamma) + \big|(\varphi,\gamma)_n\big|_{s,\varepsilon}(S) + \sum_{j=1}^{l} |\zeta_j| \Big],$$

$$\|u\|_{1,q} \le C_q \Big[\|f\|_q + \|\varphi\|_{1-1/q,q}(\Gamma) + \big\|(\varphi,\gamma)_n\big\|_{1-1/q,q}(S) + \sum_{j=1}^{l} |\zeta_j| \Big],$$

where $s = 1, \dots, k$ *and* $q \in (m, \infty)$.

PROOF. We seek a solution u in the form $u = v + w$, where v and w are solutions to the problems

$$Lv = f, \quad v\big|_\Gamma = \varphi\big|_\Gamma, \quad (v,\gamma)_n\big|_S = (\varphi,\gamma)_n\big|_S$$
$$Lw = 0, \quad w\big|_\Gamma = 0, \quad (w,\gamma)_n\big|_S = 0, \quad w(x_j) = \zeta_j - v(x_j), \quad j = 1, \dots, l,$$

respectively. For v we take the solution from Theorem 5.2 and for w we take the solution from Theorem 6.1. Then u depends linearly on the triple f, φ, $\overline{\zeta}$. By Theorem 6.1, for $s = 1, \dots, k$ and $q \in (m, \infty)$ we have the relations

$$\begin{aligned} \|w\|_{1,q} + |w|_{s,\varepsilon} &\le C_{s,q} |w|_{k,\varepsilon} \le C'_{s,q} \sum_{j=1}^{l} (|\zeta_j| + |v(x_j)|) \\ &\le C''_{s,q} \Big(\sum_{j=1}^{l} |\zeta_j| + \|v\|_{1,q} \Big) \le C'''_{s,q} \Big(\sum_{j=1}^{l} |\zeta_j| + |v|_{s,\varepsilon} \Big), \end{aligned}$$

which, together with the estimates for v, yield the required estimates. □

Now we can prove that the spaces (6.1) are infinite-dimensional.

THEOREM 6.3. *Each of the spaces* (6.1) *is infinite-dimensional.*

PROOF. The smallest space in (6.1) is $J^{k,\varepsilon}(\partial\omega, \varnothing)$. Therefore, it suffices to prove that this space is infinite-dimensional. Let $x_j \in \omega$, $j = 1, \dots, l$, $0 \ne \zeta \in \mathbb{R}^n$, and let $u_i \in C^{k,\varepsilon}(\overline{\omega}, \mathbb{R}^n)$, $i = 1, \dots, l$, be solutions to the problems

$$Lu_i(x) = 0, \quad u_i(x)\big|_{\partial\omega} = 0, \quad u_i(x_j) = \delta_{ij}\zeta.$$

The existence was proved in Theorem 6.1. Hence the $u_i(x)$ are linearly independent. Since l is arbitrary, the space $J^{k,\varepsilon}(\partial\omega, \varnothing)$ is infinite-dimensional. □

To conclude the section, we construct the projection in $C^{k,\varepsilon}(\overline{\omega}, \mathbb{R}^n)$ with the range $J^{k,\varepsilon}(\Gamma, S)$.

LEMMA 6.4. *There exists a projection* P,

$$PC^{k,\varepsilon}(\overline{\omega}, \mathbb{R}^n) = J^{k,\varepsilon}(\Gamma, S),$$

in $C^{k,\varepsilon}(\overline{\omega}, \mathbb{R}^n)$ *such that* $|Pv|_{s,\varepsilon} \le C_s |v|_{s,\varepsilon}$ *and* $\|Pv\|_{1,q} \le C_q \|v\|_{1,q}$ *for* $s = 1, \dots, k$ *and* $q \in (1, \infty)$.

PROOF. Let $v \in C^{k,\varepsilon}(\overline{\omega}, \mathbb{R}^n)$ and let $u \in C^{k,\varepsilon}(\overline{\omega}, \mathbb{R}^n)$ be the solution to the problem

$$Lu = Lv, \quad u\big|_\Gamma = v\big|_\Gamma, \quad (u,\gamma)_n\big|_S = (v,\gamma)_n\big|_S$$

from Theorem 5.2. Then $Pv \equiv v - u$ is the required projection because $P^2 = P$. □

The estimates in Lemma 6.4 show that the operator P can be extended by continuity to the spaces $\widehat{C}^{s,\varepsilon}(\overline{\omega}, \mathbb{R}^n)$ and $W^1_q(\omega, \mathbb{R}^n)$. The extended operators are bounded projections with ranges $J^{s,\varepsilon}(\Gamma, S)$ and $J_{1,q}(\Gamma, S)$ respectively.

§7. The nonnegativity condition for quadratic forms

In this section, we consider two quadratic forms

$$C[u,u] = \int_\omega b_{ij} u^i u^j \, dx,$$

$$B[u,u] = \int_\omega \left(b_{ij\alpha\beta} u^i_{x_\alpha} u^j_{x_\beta} + b_{ij\alpha} u^i_{x_\alpha} u^j + b_{ij} u^i u^j \right) dx,$$

whose coefficients are continuous functions of $x \in \overline{\omega}$. We are interested in the nonnegativity conditions for the quadratic forms C and B in two cases: the domain of the quadratic forms is the space of smooth n-dimensional vector-valued functions with compact support and the space $J^{k,\varepsilon}(\partial\omega, \varnothing)$.

THEOREM 7.1. (a) *A necessary and sufficient condition for the quadratic form $C[u,u]$ to be nonnegative definite for all smooth vector-valued functions u with compact support is that the quadratic form $b_{ij}(x)\xi_i\xi_j$ be nonnegative definite for all $x \in \overline{\omega}$ and $\xi \in \mathbb{R}^n$.*

(b) *A necessary and sufficient condition for the quadratic form $C[u,u]$ to be nonnegative definite for all $u \in J^{k,\varepsilon}(\partial\omega, \varnothing)$ is that the quadratic form $b_{ij}(x)\xi_i\xi_j$ be nonnegative definite for all $x \in \overline{\omega}$ and $\xi \in \mathbb{R}^n$.*

PROOF. (a) The sufficiency is obvious. To prove the necessity, we consider the family of functions $v_\tau(x)$ constructed in Lemma 6.1 for $\zeta_0 \in \mathbb{R}^n$ and $x_0 \in \omega$. For τ small enough the functions v_τ belong to $C^\infty_0(\omega, \mathbb{R}^n)$; therefore, $C[v_\tau, v_\tau] \geq 0$ holds for small τ. Using the formula for v_τ and making the change of variables in the integral, we find that

$$\begin{aligned} C[v_\tau, v_\tau] &= b_{ij}(x_0) \int_\omega v^i_\tau v^j_\tau \, dx + o(1) \operatorname{meas} Q_{\tau^2\tau} \\ &= \tau^{m+1} \left[b_{ij}(x_0)\zeta_{0i}\zeta_{0j} \int_{|t|\leq 1} \int_{|y|\leq 1} \alpha^2(|t|)\alpha^2(|y|) \, dt \, dy + o(1) \right] \end{aligned} \tag{7.1}$$

as $\tau \to 0$. Dividing both sides of the inequality $C[v_\tau, v_\tau] \geq 0$ by τ^{m+1} and letting τ tend to zero, we arrive at the relation $b_{ij}(x_0)\zeta_{0i}\zeta_{0j} \geq 0$ for any $x_0 \in \omega$ and $\zeta_0 \in \mathbb{R}^n$. Since the $b_{ij}(x)$ are continuous in $\overline{\omega}$, the relation remains valid at $x_0 \in \partial\omega$.

(b) The sufficiency is obvious. To prove the necessity, we have to modify the construction suggested in (a). Let $\widehat{v}_\tau \in C^{k,\varepsilon}(\overline{\omega}, \mathbb{R}^n)$ be the solution to the problem

$L\widehat{v}_\tau = Lv_\tau$, $\widehat{v}_\tau\big|_{\partial\omega} = 0$ from Theorem 5.1 for small τ. By the estimate for $\|Lv_\tau\|_q$ mentioned in Lemma 6.1,

$$\|\widehat{v}_\tau\|_{1,q} \le C_q\|Lv_\tau\|_q \le C_q\left(\operatorname{meas}^{1/q} Q_{\tau^2\tau}\right)\tau^{-1} \tag{7.2}$$

for $q \in (1,\infty)$. Since $u_\tau(x) = v_\tau(x) - \widehat{v}_\tau \in J^{k,\varepsilon}(\partial\omega, \varnothing)$, for all sufficiently small τ we obtain the inequality $C[u_\tau, u_\tau] \ge 0$, which implies

$$\int\limits_\omega b_{ij} v_\tau^i v_\tau^j\,dx + c\left(\|\widehat{v}_\tau\|_2^2 + \|v_\tau\|_2\|\widehat{v}_\tau\|_2\right) \ge 0, \quad c \ne c(\tau). \tag{7.3}$$

If $q = 2m/(m+2)$ for $m \ge 3$ and $q \ge 1$ for $m = 2$, then the embedding operator of $W_q^1(\omega)$ into $L_2(\Omega)$ is bounded. From (7.2) and the uniform boundedness of $v_\tau(x)$ in τ we obtain the estimate

$$\|\widehat{v}_\tau\|_2^2 + \|\widehat{v}_\tau\|_2\|v_\tau\|_2 \le C_q\left(\operatorname{meas} Q_{\tau^2\tau}\right) R(R+1),$$
$$R = \left(\operatorname{meas} Q_{\tau^2\tau}\right)^{(2-q)/(2q)}\tau^{-1} = \left|S_{m-1}\right|^{(2-q)/(2q)}\tau^{-1+(m+1)(2-q)/(2q)}.$$

The exponent τ in the expression for R is positive if $q < 2(m+1)/(m+3)$. This inequality is satisfied by $q = 2m/(m+2)$, as well as by any $q \in [1, 6/5)$ if $m = 2$. Combining the result with (7.3), we conclude that

$$\int\limits_\omega b_{ij} v_\tau^i v_\tau^j\,dx + o(1)\operatorname{meas} Q_{\tau^2\tau} \ge 0 \tag{7.4}$$

as $\tau \to 0$. We divide both sides of (7.4) by $\operatorname{meas} Q_{\tau^2\tau}$ and pass to the limit as $\tau \to 0$. Taking into account (7.1), we obtain the inequality $b_{ij}(x_0)\zeta_{0i}\zeta_{0j} \ge 0$ for all $x_0 \in \omega$, $\zeta_0 \in \mathbb{R}^n$. Since the $b_{ij}(x)$ are continuous in $\overline{\omega}$, the inequality remains valid at $x_0 \in \partial\omega$. □

Theorem 7.1 shows that the requirement that $Lu = 0$ in the domain of the form $C[u,u]$ does no affect the criterion for nonnegativity of a quadratic form. A quite different situation arises in the case of $B[u,u]$.

THEOREM 7.2. (a) *The inequality* $b_{ij\alpha\beta}(x)\xi_i\xi_j\lambda_\alpha\lambda_\beta \ge 0$ *for all* $x \in \overline{\omega}$, $\xi \in \mathbb{R}^n$, $\lambda \in \mathbb{R}^m$ *is a necessary condition for the quadratic form* $B[u,u]$ *to be nonnegative for all smooth* n*-dimensional functions* u *with compact support.*

(b) *The inequality* $b_{ij\alpha\beta}(x)\xi_i\xi_j\lambda_\alpha\lambda_\beta \ge 0$ *for all* $x \in \overline{\omega}$, $\xi \in \mathbb{R}^n$, $\lambda \in \mathbb{R}^m$ *is not necessary for the quadratic form* $B[u,u]$ *to be nonnegative for all* $u \in J^{k,\varepsilon}(\partial\omega,\varnothing)$.

(c) *The inequality* $b_{ij\alpha\beta}(x)\xi_i\xi_j\lambda_\alpha\lambda_\beta \ge 0$ *for all* $x \in \overline{\omega}$ *and those vectors* $\xi \in \mathbb{R}^n$, $\lambda \in \mathbb{R}^m$, *which satisfy the orthogonality condition at* $x \in \overline{\omega}$, *is a necessary condition for the quadratic form* $B[u,u]$ *to be nonnegative definite for all* $u \in J^{k,\varepsilon}(\partial\omega,\varnothing)$.

PROOF. (a) For the functions $v_{l\tau}(x)$ constructed in Lemma 6.1 and their derivatives (cf. §6) the following estimates hold:

$$\begin{aligned} &|v_{l\tau}(x)| \le C, \qquad \left|(v_{l\tau}(x))_{x_\alpha}\right| \le C\left(l^{-1} + \tau^{-1}\right), \\ &\left|(v_{l\tau}^i(x))_{x_\beta} - l^{-1}\zeta^i\lambda_\beta\dot{\alpha}(|t|/l)\alpha(|y|/\tau)\operatorname{sgn} t\right| \le C\tau^{-1}, \end{aligned} \tag{7.5}$$

where $0 < l \le \tau \le 1$, $x \in Q_{l\tau}$, and $C \ne C(l,\tau)$ is a constant. We assume that $x_0 \in \omega$. In this case, for sufficiently small τ the functions $v_{l\tau}(x)$ have compact support in ω. In view of (7.5) and the nonnegativity of the quadratic form $B[v_{l\tau}, v_{l\tau}]$, we get

$$c(l^{-1}+\tau^{-1})\operatorname{meas} Q_{l\tau} + \int\limits_{Q_{l\tau}} b_{ij\gamma\beta}\left(v^i_{l\tau}\right)_{x_\gamma}\left(v^j_{l\tau}\right)_{x_\beta}dx \geq 0. \tag{7.6}$$

To estimate the second term in (7.6) from above, we use the third inequality in (7.5):

$$\begin{aligned}\int\limits_{Q_{l\tau}} & b_{ij\gamma\beta}\left(v^i_{l\tau}\right)_{x_\gamma}\left(v^j_{l\tau}\right)_{x_\beta}dx \\ &\leq l^{-2}\lambda_\gamma\lambda_\beta\int\limits_{Q_{l\tau}} b_{ij\gamma\beta}\dot{\alpha}^2(|t|/l)\alpha^2(|y|/\tau)\zeta^i(x)\zeta^j(x)\,dx \\ &\quad + C\operatorname{meas} Q_{l\tau}[\tau^{-1}l^{-1}+\tau^{-2}].\end{aligned}$$

From (7.6) and the last inequality it follows that

$$\begin{aligned}& c\left(l^{-1}+\tau^{-1}+l^{-1}\tau^{-1}+\tau^{-2}\right)\operatorname{meas} Q_{l\tau} \\ &\quad + l^{-2}\lambda_\gamma\lambda_\beta\int\limits_{|t|\leq l}\int\limits_{|y|\leq\tau} b_{ij\gamma\beta}(x)\dot{\alpha}^2(|t|/l)\alpha^2(|y|/\tau)\zeta^i(x)\zeta^j(x)\,dx \geq 0,\end{aligned}$$

where $x = x_0 + t\lambda + y$. Changing coordinates $tl^{-1} \to t$, $y\tau^{-1} \to y$ in the integral and dividing both sides of the result by $l^2(l\tau^{m-1})$, we arrive at the inequality

$$\begin{aligned}&\lambda_\gamma\lambda_\beta\int\limits_{|t|\leq 1}\int\limits_{|y|\leq 1} b_{ij\gamma\beta}(x)\dot{\alpha}^2(|t|)\alpha^2(|y|)\zeta^i(x)\zeta^j(x)\,dx \\ &\quad + c\left(l + l^2\tau^{-1} + l\tau^{-1} + l^2\tau^{-2}\right) \geq 0,\end{aligned}$$

where $x = x_0 + tl\lambda + \tau y$. We take l and τ to zero so as to satisfy the conditions $0 < l \leq \tau \leq 1$ and $l\tau^{-1} \to 0$ (these conditions hold a fortiori if $0 < \tau \leq 1$ and $l = \tau^2$). After the passage to the limit we obtain the inequality $b_{ij\alpha\beta}(x_0)\zeta^i(x_0)\zeta^j(x_0)\lambda_\alpha\lambda_\beta \geq 0$. Since no conditions on $\zeta(x_0)$ were required, $\zeta(x_0)$ is an arbitrary vector in $\mathbb{R}^n$. The single condition $|\lambda| = 1$ was imposed on $\lambda \in \mathbb{R}^m$. But if the inequality obtained is valid for a unit vector λ, then it remains valid for an arbitrary vector λ. Hence the inequality holds at an arbitrary point $x_0 \in \omega_0$. By the continuity of $b_{ij\alpha\beta}(x)$, it remains valid at $x_0 \in \partial\omega$.

(b) We define the quadratic form $B[u,u]$ by the equality

$$B[u,u] = -\int_\omega (Lu)^2\,dx.$$

It can be written in a standard form; moreover, $b_{ij\alpha\beta}(x) = -A_{i\alpha}(x)A_{j\beta}(x)$. Hence $b_{ij\alpha\beta}(x)\xi_i\xi_j\lambda_\alpha\lambda_\beta = -(A^*(x)\xi,\lambda)^2_m$. On the space $J^{k\varepsilon}(\partial\omega,\varnothing)$, the quadratic form $B[u,u]$ is identically zero, but the inequality $b_{ij\alpha\beta}(x)\xi_i\xi_j\lambda_\alpha\lambda_\beta \geq 0$ fails for arbitrary $x \in \overline{\omega}$, $\xi \in \mathbb{R}^n$, and $\lambda \in \mathbb{R}^m$. Indeed, fix $x \in \overline{\omega}$ and take $\xi = A(x)\lambda$, $\lambda \neq 0$. Then $b_{ij\alpha\beta}(x)\xi_i\xi_j\lambda_\alpha\lambda_\beta = -(G(x)\lambda,\lambda)^2_m < 0$, since the nonnegative matrix G is nonsingular.

(c) Let $x_0 \in \omega$ and let $v_{l\tau}(x)$ be the family of functions constructed in Lemma 6.1. We impose the orthogonality condition $(A(x_0)\lambda,\xi)_n = 0$ for $\xi = \zeta(x_0)$ and λ.

Then the estimate from Lemma 6.1 is valid for $\|Lv_{l\tau}\|_q$. Let τ be small enough and let $\widehat{v}_{l\tau}(x)$ be the solution to the problem $L\widehat{v}_{l\tau}(x) = Lv_{l\tau}(x)$, $\widehat{v}_{l\tau}\big|_{\partial\omega} = v_{l\tau}\big|_{\partial\omega} = 0$ from Theorem 5.1. Then $u_{l\tau}(x) = v_{l\tau}(x) - \widehat{v}_{l\tau} \in J^{k,\varepsilon}(\partial\omega, \varnothing)$. Furthermore, we have the inequality $B[u_{l\tau}, u_{l\tau}] \geq 0$, which implies the relation

$$B[v_{l\tau}, v_{l\tau}] + c\left[\|\widehat{v}_{l\tau}\|_{1,2}^2 + \|v_{l\tau}\|_{1,2}\|\widehat{v}_{l\tau}\|_{1,2}\right] \geq 0. \tag{7.7}$$

Using the estimate for the solution $\widehat{v}_{l\tau}$ from Theorem 5.1, the estimate for the $L_2(\omega)$-norm in $Lv_{l\tau}$ from Lemma 6.1, and the estimates for $v_{l\tau}$ and its derivatives (cf. (7.5)), we find that

$$\begin{aligned}\|\widehat{v}_{l\tau}\|_{1,2}^2 + \|v_{l\tau}\|_{1,2}\|\widehat{v}_{l\tau}\|_{1,2} &\leq c\left[\|Lv_{l\tau}\|_{1,2}^2 + \|Lv_{l\tau}\|_2\|v_{l\tau}\|_{1,2}\right] \\ &\leq c(\operatorname{meas} Q_{l\tau})\left[(\tau l^{-1} + \tau^{-1})^2 + (\tau l^{-1} + \tau^{-1})(l^{-1} + \tau^{-1})\right].\end{aligned}$$

Then (7.7) implies the inequality

$$\frac{B[v_{l\tau}, v_{l\tau}]}{l^{-2}(l\tau^{m-1})} + c\left[(\tau + l\tau^{-1})^2 + (\tau + l\tau^{-1})(1 + l\tau^{-1})\right] \geq 0.$$

Passing to the limit in the same way as in the proof of (a), we obtain the inequality $b_{ij\alpha\beta}(x_0)\xi_i\xi_j\lambda_\alpha\lambda_\beta \geq 0$. The latter is valid for $x_0 \in \omega$, $\xi \in \mathbb{R}^n$, and $\lambda \in \mathbb{R}^m$ connected by the relation $(A(x_0)\lambda, \xi)_n = 0$. To complete the proof, it remains to show that a point x_0 may belong to $\partial\omega$. For $\lambda = 0$ the inequality is obvious. Let $\lambda \neq 0$. Then $A(x_0)\lambda \neq 0$ for all $x_0 \in \overline{\omega}$. Therefore, the set of all vectors $\xi \in \mathbb{R}^n$ such that $(A(x_0)\lambda, \xi)_n = 0$ can be written as follows: $\xi = \zeta - (A(x_0)\lambda, \zeta)_n|A(x_0)\lambda|^{-2}A(x_0)\lambda$, where ζ is an arbitrary vector of $\mathbb{R}^n$. For such $\xi = \xi(x_0)$ the inequality $b_{ij\alpha\beta}(x_0)\xi_i(x_0)\xi_j(x_0)\lambda_\alpha\lambda_\beta \geq 0$ holds at $x_0 \in \omega$. Since the coefficients $b_{ij\alpha\beta}(x_0)$ and vectors $\xi(x_0)$ are continuous with respect to x_0 in $\overline{\omega}$, the same inequality holds for $x_0 \in \partial\omega$. □

As an application of the above results, we study the question of restoring the operator L from its kernel, i.e., from the space $J^{k,\varepsilon}(\partial\omega, \varnothing)$. We begin with the following algebraic lemma.

LEMMA 7.1. *Let $A, B : \mathbb{R}^m \to \mathbb{R}^n$ be linear operators and let the $m \times m$ matrix $G = A^*A$ be nonsingular. Suppose that for all $\lambda \in \mathbb{R}^m$ and $\xi \in \mathbb{R}^n$ satisfying the relation*

$$(A\lambda, \xi)_n = 0 \tag{7.8}$$

the following equality holds:

$$(B\lambda, \xi)_n = 0. \tag{7.9}$$

Then $B = (\operatorname{tr} A^*B)(\operatorname{tr} A^*A)^{-1}A$.

PROOF. Since the matrix G is nonsingular, for $\lambda \in \mathbb{R}^m$ different from zero we have $A\lambda \neq 0$. Therefore, the vectors $\xi = \zeta - (\zeta, A\lambda)_n(A\lambda, A\lambda)_n^{-1}A\lambda$ and $A\lambda$ are orthogonal in $\mathbb{R}^n$ for any $\zeta \in \mathbb{R}^n$, i.e., the vectors λ and ξ satisfy (7.8). By the assumption of the lemma, they also satisfy (7.9), which can be written in the form

$$(B\lambda, \zeta)_n = c(\lambda)(A\lambda, \zeta)_n, \quad c(\lambda) = (A\lambda, B\lambda)_n(A\lambda, A\lambda)_n^{-1}.$$

Since $\zeta \in \mathbb{R}^n$ is arbitrary, this relation is equivalent to the equality

$$B\lambda = c(\lambda)A\lambda, \quad \lambda \in \mathbb{R}^m, \quad \lambda \neq 0. \tag{7.10}$$

We prove that $c(\lambda)$ is independent of λ. If λ_1 and λ_2 are linearly dependent, then the equality $c(\lambda_1) = c(\lambda_2)$ follows from the definition of $c(\lambda)$ because λ_1 and λ_2 are proportional. Let λ_1 and λ_2 be linearly independent. From (7.10) we obtain the relations

$$Q\lambda = c(\lambda)\lambda, \qquad Q = G^{-1}A^*B. \tag{7.11}$$

Write (7.11) for $\lambda = \lambda_1$, $\lambda = \lambda_2$, $\lambda = \lambda_1 + \lambda_2$. We add the first two equalities and subtract the third one from the sum. We obtain

$$[c(\lambda_1) - c(\lambda_1 + \lambda_2)]\,\lambda_1 + [c(\lambda_2) - c(\lambda_1 + \lambda_2)]\,\lambda_2 = 0.$$

Since λ_1 and λ_2 are linearly independent, we get $c(\lambda_1) = c(\lambda_1+\lambda_2) = c(\lambda_2)$. Thus, $c(\lambda) \equiv c = \text{const}$ for $\lambda \neq 0$. Therefore, (7.10) is equivalent to the equality $B = cA$. Multiplying both sides of this equality by A^* and taking the trace, we find the expression for c. □

We introduce the operator

$$Mu = \operatorname{div} B^*(x)u + (b(x), u)_n, \tag{7.12}$$

where $B(x)$ is a matrix-valued function with components $B_{ij}(x)$, $i = 1, \dots, n$, $j = 1, \dots, m$, of class $C^1(\overline{\omega})$, $b(x)$ is a vector-valued function with coordinates $b^i(x)$, $i = 1, \dots, n$, of class $C(\overline{\omega})$, and $u(x)$ is an n-dimensional vector-valued function. It is obvious that (7.12) presents a general form of scalar first order operator (with coefficients smooth enough) acting on the class of n-dimensional vector-valued functions. Our goal is to prove that any first order operator whose kernel is not less than the kernel of L, differs from L only by a functional factor.

THEOREM 7.3. *Let $Mu = 0$ for all $u \in J^{k,\varepsilon}(\partial\omega, \varnothing)$. Then for $x \in \overline{\omega}$ we have*

$$M = [\operatorname{tr} A^*(x)B(x)][\operatorname{tr} A^*(x)A(x)]^{-1}L.$$

PROOF. On $J^{k,\varepsilon}(\partial\omega, \varnothing)$, we consider two quadratic forms

$$K_\pm[u,u] = \int_\omega \left[\pm(Mu)^2\right]\, dx. \tag{7.13}$$

It is obvious that $(Mu)^2 = b_{ij\alpha\beta}(x)u^i_{x_\alpha}u^j_{x_\beta} + b_{ij\alpha}(x)u^i_{x_\alpha}u^j + b_{ij}(x)u^iu^j$; moreover, the coefficients $b(x)$ with different subscripts are continuous in $\overline{\omega}$ and the relation $b_{ij\alpha\beta}(x)\xi_i\xi_j\lambda_\alpha\lambda_\beta = (B(x)\lambda, \xi)^2_n$ holds for any $\xi \in \mathbb{R}^n$, $\lambda \in \mathbb{R}^m$, where $B(x)$ is the matrix corresponding to the operator M. Since $Mu = 0$ for $u \in J^{k,\varepsilon}(\partial\omega, \varnothing)$, both the quadratic forms $K_\pm[u,u]$ are identically zero. Therefore, for each point $x \in \overline{\omega}$ and vectors $\xi \in \mathbb{R}^n$, $\lambda \in \mathbb{R}^m$ such that $(A(x)\lambda, \xi)_n = 0$, we have $(B(x)\lambda, \xi)_n = 0$. Hence Lemma 7.1 yields

$$B(x) = c(x)A(x), \quad c(x) = [\operatorname{tr} A^*(x)B(x)]\,[\operatorname{tr} A^*(x)A(x)]^{-1}. \tag{7.14}$$

Differentiating and using (7.14), we write out two equalities for $u \in J^{k,\varepsilon}(\partial\omega, \varnothing)$:

$$0 = c(x)Lu = B_{ij}(x)u^i_{x_j} + c(x)\left[a^i(x) + (A_{ij}(x))_{x_j}\right]u^i,$$
$$0 = Mu = B_{ij}(x)u^i_{x_j} + \left[b^i(x) + (B_{ij}(x))_{x_j}\right]u^i.$$

Subtracting one from other, we obtain

$$(7.15)\qquad \begin{gathered}(p(x), u(x))_n = 0, \quad x \in \overline{\omega}, \\ p^i(x) = c(x)\left[a^i(x) + (A_{ij}(x))_{x_j}\right] - \left[b^i(x) + (B_{ij}(x))_{x_j}\right], \; i = 1, \dots, n,\end{gathered}$$

for $u \in J^{k,\varepsilon}(\partial\omega, \varnothing)$. By Theorem 6.1, for every $x_0 \in \omega$ we can indicate a function $u \in J^{k,\varepsilon}(\partial\omega, \varnothing)$ such that $u(x_0) = p(x_0)$. Since (7.15) is valid for all such functions $u(x)$, we have $p(x) \equiv 0$ in ω, and in $\overline{\omega}$ in view of the continuity of $p(x)$. With the help of (7.14) for $B(x)$, we rewrite the equality $p(x) = 0$ in the form

$$c(x)\left[a^i(x) + (A_{ij}(x))_{x_j}\right] = \left[b^i(x) + (c(x)A_{ij}(x))_{x_j}\right], \quad i = 1, \dots, n,$$

whence $b(x) = c(x)a(x) - A(x)\nabla c(x)$. Therefore, for any vector-valued function $v \in C^1(\overline{\omega}, \mathbb{R}^n)$ we have

$$\begin{aligned}Mv &= \operatorname{div} B^* v + (b, v)_n = \operatorname{div}(c(x)A^* v) + (c(x)a - A\nabla c(x), v)_n \\ &= c(x)\left[\operatorname{div} A^* v + (a, v)_n\right] = c(x)Lv,\end{aligned}$$

which is equivalent to the required relation because v is arbitrary and (7.14) holds for $c(x)$. □

§8. Elements of the kernel of L with fixed values

Let $P(x)$ be an orthogonal projection in $\mathbb{R}^n$ for $x \in \overline{\omega}$ with matrix-valued coefficients of class $C^{k,\varepsilon}(\overline{\omega})$. We consider the problem

$$(8.1)\qquad Lu = 0, \quad u\big|_{\partial\omega} = 0, \quad P(x)u(x) = \varphi(x),$$

where $\varphi \in C^{k,\varepsilon}(\overline{\omega}, \mathbb{R}^n)$, $\varphi\big|_{\partial\omega} = 0$, $P(x)\varphi(x) \equiv \varphi(x)$, and $u \in C^{k,\varepsilon}(\overline{\omega}, \mathbb{R}^n)$ is an unknown function. A solution to this problem is an element of the space $J^{k,\varepsilon}(\partial\omega, \varnothing)$ with a fixed value in the subspace $P(x)\mathbb{R}^n$ for every $x \in \overline{\omega}$. We seek a solution u in the form

$$(8.2)\qquad u = (1 - P)v + \varphi, \quad v \in C^{k\varepsilon}(\overline{\omega}, \mathbb{R}^n), \qquad v\big|_{\partial\omega} = 0.$$

Then the problem (8.1) takes the form

$$(8.3)\qquad L_P v = -L\varphi, \qquad v\big|_{\partial\omega} = 0.$$

The structure of the operator L_P is the same as that of the operator L with A and a replaced by $A_P = (1 - P)A$ and $a_P = (1 - P)a$. Let $G_P = A_P^*(x)A_P(x)$.

LEMMA 8.1. *The relation* $\det G_P(x) \neq 0$ *for* $x \in \overline{\omega}$ *is equivalent to the equality* $RA(x) \cap RP(x) = \{0\}$.

PROOF. The relation $\det G_P(x) \neq 0$ is equivalent to the fact that any solution $\xi \in \mathbb{R}^m$ to $G_P(x)\xi = 0$ is zero. Since the symmetric matrix $G_P(x)$ is nonnegative, the last vector equality is equivalent to the scalar equality $(G_P(x)\xi, \xi)_m = 0$. Since $(G_P(x)\xi, \xi)_m = |(1 - P(x))A(x)\xi|^2$, the condition $\det G_P(x) \neq 0$ is equivalent to the condition that any solution ξ to $A(x)\xi = P(x)A(x)\xi$ is zero. Since the matrix $G(x)$ is nonsingular in $\overline{\omega}$, we have $A(x)\xi \neq 0$ if $\xi \neq 0$. Therefore, any solution to the last equation is zero if and only if $RP(x) \cap RA(x) = 0$. □

LEMMA 8.2. *Let $RP(x) \cap RA(x) = \{0\}$ for all $x \in \overline{\omega}$. The problem $L_P^* q = 0$, $q \in C^1(\overline{\omega})$, has a nonzero solution if and only if the coefficient $a(x)$ admits a representation*

$$a(x) = A(x)\nabla\psi(x) + h(x), \tag{8.4}$$

where

$$\psi \in C^{k+1,\varepsilon}(\overline{\omega}), \quad h \in C^{k,\varepsilon}(\overline{\omega}, \mathbb{R}^n), \quad P(x)h(x) \equiv h(x).$$

If such a representation exists, any solution $q(x)$ is proportional to $q^* = \exp\psi$.

PROOF. By Theorem 2.1, a criterion for nontrivial solvability of the problem is the representability of $a_P(x)$ in the form $a_P(x) = A_P(x)\nabla\psi(x)$ with some function $\psi \in C^{k+1,\varepsilon}(\overline{\omega})$, which implies (8.4) and the formula for the solution. □

We now turn to the problem (8.1).

THEOREM 8.1. *We assume that $RA(x) \cap RP(x) = \{0\}$ for all $x \in \overline{\omega}$.*

(a) *If $a(x)$ is not represented in the form (8.4) with any functions $\psi(x)$ and $h(x)$, then there exists a solution $u \in C^{k,\varepsilon}(\overline{\omega}, \mathbb{R}^n)$ to (8.1) that depends linearly on φ and satisfies the following inequalities for $s = 1, \dots, k$ and $q \in (1, \infty)$:*

$$|u|_{s,\varepsilon} \le C_s|\varphi|_{s,\varepsilon}, \quad \|u\|_{1,q} \le C_q\|\varphi\|_{1,q}. \tag{8.5}$$

(b) *If $a(x)$ is represented in the form (8.4), then the equality*

$$\int_\omega e^\psi (h, \varphi)_n \, dx = 0 \tag{8.6}$$

is a necessary and sufficient condition for the solvability of the problem (8.1). If (8.6) holds, then there exists a solution possessing the properties indicated in (a).

PROOF. (a) The assertion follows from Theorem 5.1(a) and Lemma 8.2 applied to the problem (8.3).

(b) To prove the necessity, we multiply (8.1) by $\exp\psi$, integrate by parts, and use (8.4). As a result, we obtain

$$0 = \int_\omega e^\psi Lu \, dx = \int_\omega (u, L^* e^\psi)_n \, dx = \int_\omega e^\psi (u, h)_n \, dx = \int_\omega e^\psi (\varphi, h)_n \, dx.$$

To prove the sufficiency, we apply Theorem 5.1(b) and Lemma 8.2 to the problem (8.3). Since

$$\int_\omega e^\psi L\varphi \, dx = \int_\omega (L^* e^\psi, \varphi)_n \, dx = \int_\omega e^\psi (\varphi, h)_n \, dx,$$

the consistency condition (8.6) coincides with the consistency condition (5.2) for the problem (8.3). □

The following problem is more general than (8.1):

$$Lu = f, \quad u\big|_{\partial\omega} = 0, \quad P(x)u(x) = \varphi(x), \tag{8.7}$$

where $\varphi(x)$ satisfies the same conditions as above and $f \in C^{k-1,\varepsilon}(\overline{\omega})$.

THEOREM 8.2. *Suppose that $RA(x) \cap RP(x) = \{0\}$ for all $x \in \overline{\omega}$.*

(a) *If $a(x)$ is not represented in the form* (8.4) *with any functions $\psi(x)$ and $h(x)$, then there exists a solution $u \in C^{k,\varepsilon}(\overline{\omega}, \mathbb{R}^n)$ to the problem* (8.7) *that depends linearly on f, φ and satisfies the inequalities*

$$\begin{aligned} |u|_{s,\varepsilon} &\le C_s\left[|f|_{s-1,\varepsilon} + |\varphi|_{s,\varepsilon}\right], \\ \|u\|_{1,q} &\le C_q\left[\|f\|_q + \|\varphi\|_{1,q}\right], \end{aligned} \tag{8.8}$$

where $s = 1, \ldots, k$ and $q \in (1, \infty)$.

(b) *If $a(x)$ is represented in the form* (8.4), *then the equality*

$$\int_\omega e^\psi\left[(h, \varphi)_n - f\right] dx = 0 \tag{8.9}$$

yields a criterion for solvability of the problem (8.7). *If* (8.9) *holds, then there exists a solution u possessing the properties indicated in* (a).

PROOF. The proof is similar to that of Theorem 8.1. □

REMARK 8.1. Let $l \in C^{k,\varepsilon}(\overline{\omega}, \mathbb{R}^n)$, $|l(x)| \equiv 1$. We consider a special case of the orthogonal projection $P(x) = l(x)(l(x), \cdot)_n$ on the field $l(x)$. In this case, the nondegeneracy condition for the matrix $G_P(x)$ and the representation (8.4) can be rewritten in the form

$$\begin{gathered} l(x) \notin RA(x), \quad x \in \overline{\omega}, \quad a(x) = A(x)\nabla\psi(x) + \alpha(x)l(x), \\ \varphi \in C^{k,\varepsilon}(\overline{\omega}), \quad \alpha \in C^{k,\varepsilon}(\overline{\omega}), \end{gathered}$$

and the condition (8.9) is reduced to the equality $\int_\omega e^\psi[\alpha(l, \varphi)_n - f]\, dx = 0$.

§9. Decomposition of vector fields

Let $\omega \subset \mathbb{R}^m$ be a bounded domain with boundary $\partial\omega \in C^{k+1,\varepsilon}$, $k \ge 1$, $\varepsilon \in (0, 1]$. We decompose a vector field f of class $C^{k,\varepsilon}(\overline{\omega}, \mathbb{R}^n)$ into the sum

$$f = u + L^*p, \tag{9.1}$$

where u and p depend linearly on f and satisfy the conditions

$$\begin{gathered} u \in C^{k,\varepsilon}(\overline{\omega}, \mathbb{R}^n), \quad Lu = 0, \quad (u - f, A\nu)_n\big|_{\partial\omega} = 0, \quad p \in C^{k+1,\varepsilon}(\overline{\omega}), \\ |L^*p|_{s,\varepsilon} + |u|_{s,\varepsilon} \le C_s|f|_{s,\varepsilon}, \quad s = 1, \ldots, k, \\ \|L^*p\|_{1,q} + \|u\|_{1,q} \le C_q\|f\|_{1,q}, \quad q \in (1, \infty), \\ \|L^*p\|_q + \|u\|_q \le C_q\|f\|_q, \quad q \in (1, \infty). \end{gathered} \tag{9.2}$$

THEOREM 9.1. *There exists a decomposition* (9.1), (9.2). *The functions u and L^*p are uniquely determined.*

PROOF. To prove the uniqueness, we note that the equality $u = -L^*p$, where

$$u \in C^{k,\varepsilon}(\overline{\omega}, \mathbb{R}^n), \quad Lu = 0, \quad (u, A\nu)_n\big|_{\partial\omega} = 0, \quad p \in C^{k+1,\varepsilon}(\overline{\omega}),$$

holds only if $u \equiv 0$, because its terms are orthogonal in $L_2(\omega, \mathbb{R}^n)$.

We pass to the proof of existence. We seek a function $p(x)$ from the condition

$$L(L^*p - f) = 0, \qquad (L^*p - f, A\nu)_n\big|_{\partial\omega} = 0. \tag{9.3}$$

Since

$$Tp \equiv LL^*p = -\left[G_{ij}p_{x_j}\right]_{x_i} + (La)p,$$
$$(L^*p, A\nu)_n = -G_{ij}p_{x_j}\nu_i + (a, A\nu)_n p$$

and the matrix $G(x) = A^*(x)A(x)$ is positive definite, the conditions (9.3) can be regarded as the third boundary-value problem for a second order elliptic operator T with coefficients $G_{ij} \in C^{k,\varepsilon}(\overline{\omega})$, $(La) \in C^{k-1,\varepsilon}(\overline{\omega})$, $(a, A\nu)_n \in C^{k,\varepsilon}(\partial\omega)$:

$$(9.4) \qquad Tp = Lf, \quad x \in \overline{\omega}, \qquad -G_{ij}p_{x_j}\nu_i + (a, A\nu)_n p = (f, A\nu)_n, \quad x \in \partial\omega;$$

moreover, $Lf \in C^{k-1,\varepsilon}(\overline{\omega})$ and $(f, A\nu)_n \in C^{k,\varepsilon}(\partial\omega)$.

Let p_0 be a solution to the homogeneous ($f \equiv 0$) problem (9.4) or, equivalently, to the homogeneous problem (9.3). Multiplying (9.3) by p_0, integrating the result by parts, and taking into account the homogeneous boundary condition, we conclude that $L^*p_0 = 0$. If the problem $L^*p_0 = 0$ has only the zero solution, then the homogeneous problem (9.4) also has only the zero solution and the problem (9.3) is uniquely solvable. Let the equation $L^*p_0 = 0$ have a nonzero solution. Then the solution is proportional to p^*, and the criterion for solvability of the problem (9.4) for a given f can be described by the equality

$$\int\limits_{\omega} (Lf)p^*\,dx = \int\limits_{\partial\omega} (f, A\nu)_n p^*\,dS.$$

Since $L^*p^* = 0$, the last equality is valid because of the integration by parts formula. Thus, a solution to the problem (9.4) exists, but it is determined up to a summand cp^*. We eliminate the arbitrariness by imposing the condition

$$(9.5) \qquad \Phi(p) \equiv \left| \int\limits_{\omega} p(x)p^*(x)\,dx \right| = 0$$

on a solution. The solution depends linearly on the right-hand side. In both cases, $p \in C^{k+1,\varepsilon}(\overline{\omega})$ satisfies the estimates

$$|p|_{s+1,\varepsilon} \le C_s\left[|Lf|_{s-1,\varepsilon}(\overline{\omega}) + |(f, A\nu)_n|_{s,\varepsilon}(\partial\omega)\right], \quad s = 1, \dots, k,$$
$$\|p\|_{2,q} \le C_q\left[\|Lf\|_q(\omega) + \|(f, A\nu)_n\|_{1-1/q,q}(\partial\omega)\right], \quad q \in (1, \infty).$$

Since

$$|(f, A\nu)_n|_{s,\varepsilon}(\partial\omega) \le C|f|_{s,\varepsilon}(\partial\omega) \le C'|f|_{s,\varepsilon}(\overline{\omega}),$$
$$\|(f, A\nu)_n\|_{1-1/q,q}(\partial\omega) \le C\|f\|_{1-1/q,q}(\partial\omega) \le C'\|f\|_{1,q}(\omega),$$

these inequalities lead to the following estimates:

$$|L^*p|_{s,\varepsilon} \le C_s|f|_{s,\varepsilon}, \qquad \|L^*p\|_{1,q} \le C_q\|f\|_{1,q}.$$

The function $u \equiv f - L^*p$ satisfies the first condition from (9.2). The above estimates provide the following estimates for u:

$$|u|_{s,\varepsilon} \le C_s|f|_{s,\varepsilon}, \qquad \|u\|_{1,q} \le C_q\|f\|_{1,q}.$$

Thus, the functions u and p just constructed yield the decomposition (9.1) and satisfy the first three conditions in (9.2). To prove the third estimate in (9.2), it suffices to establish the inequality

$$\|p\|_{1,q} \leq C_q \|f\|_q, \qquad q \in (1, \infty), \tag{9.6}$$

for our solution to (9.4) and repeat the above arguments. Let us prove (9.6) for $q = 2$. Multiplying (9.3) by $p(x)$ and integrating by parts, we find that

$$\int_\omega |L^* p|^2 \, dx = \int_\omega (f, L^* p)_n \, dx \leq \|f\|_2 \|L^* p\|_2.$$

This estimate implies $\|L^* p\|_2 \leq \|f\|_2$. To obtain (9.6) for $q = 2$, it remains to recall Lemma 2.3 or Lemma 2.4 and (9.5). For an arbitrary q the proof of (9.6) is based on the equality

$$\int_\omega \left[G_{ij} p_{x_i} h_{x_j} + (La) ph \right] dx - \int_{\partial\omega} (A\nu, a)_n ph \, dS = \int_\omega (f, L^* h)_n \, dx \tag{9.7}$$

for any $h \in C^1(\overline{\omega})$, which is equivalent to the problem (9.4). We divide the proof into several steps formulated as lemmas. □

LEMMA 9.1. *Let $u \in W_q^1(D)$, $f \in L_q(D, \mathbb{R}^m)$, and $g \in L_q(D)$, with $q > 1$ and $D = B_R(0)$, vanish in some neighborhood of ∂D and satisfy the equality*

$$\int_D [(\nabla u + f, \nabla h)_m + gh] \, dx = 0 \tag{9.8}$$

for any $h \in C_0^\infty(D)$. Then u satisfies the estimate

$$\|\nabla u\|_q \leq C_q [\|f\|_q + \|g\|_q]. \tag{9.9}$$

PROOF. Since the functions u, f, and g are compactly supported, (9.8) also holds for $h \in C^\infty(\overline{D})$. Therefore, in (9.8), instead of h we can take its average h_ρ, where $\rho \neq \rho(h)$ is small enough. From (9.8) and the properties of averages it follows that u_ρ, f_ρ, and g_ρ also satisfy (9.8). The last assertion is equivalent to the equality

$$\Delta u_\rho = g_\rho - \operatorname{div} f_\rho,$$

where $x \in D$, $u_\rho, f_\rho, g_\rho \in C_0^\infty(D)$. From the representation of $u_\rho(x)$ in terms of the Newton potential we easily obtain (cf. [**21**])

$$(u_\rho)_{x_i}(x) = -|S_1| f_\rho^i(x) + \int_D g_\rho(y) E_{y_i}(x - y) \, dy + \int_D f_\rho^j(y) E_{y_i y_j}(x - y) \, dy,$$

where $|S_1|$ denotes the area of the unit sphere in $\mathbb{R}^m$ and E is the fundamental solution to the Laplace operator. The last integral is thought of as a singular one. Singular integral operators and integral operators with weak singularities are bounded in L_q, $q > 1$ (cf. [**21**], [**36**]). Hence

$$\left\|(\nabla u)_\rho\right\|_q \leq C_q \left[\left\|f_\rho\right\|_q + \left\|g_\rho\right\|_q \right].$$

Taking ρ to zero, we obtain (9.9). □

LEMMA 9.2. *We assume that functions u, f, and g belong to the spaces indicated in Lemma 9.1, but over domain $D = B_R(0) \cap \mathbb{R}^m_+$, vanish in some neighborhood of the spherical part of ∂D, and satisfy (9.8) for all h that are infinitely differentiable in $\overline{D}$ and vanish in a neighborhood of the spherical part of ∂D. Then u satisfies (9.9).*

PROOF. We extend u, $f_1, \dots, f_{m-1}$, and g as even functions and f as an odd function for negative x_m. The extended functions will be denoted by $\widehat{u}$, $\widehat{f}$, and $\widehat{g}$ respectively. They belong to the spaces $W^1_q(B_R(0))$, $L_q(B_R(0), \mathbb{R}^m)$, and $L_q(B_R(0))$ and have compact support. We verify that $\widehat{u}$, $\widehat{f}$, and $\widehat{g}$ satisfy (9.8) in $D = B_R(0)$. We decompose $h \in C_0^\infty(B_R(0))$ into the sum of odd and even components:

$$\begin{aligned} h(x', x_m) &= [h(x', x_m) + h(x', -x_m)]/2 + [h(x', x_m) - h(x', -x_m)]/2 \\ &\equiv h_1(x) + h_2(x), \end{aligned}$$

where $x' = (x_1, \dots, x_{m-1})$. The functions $\widehat{u}$, $\widehat{f}$, $\widehat{g}$, and h_2 satisfy (9.8) in $D = B_R(0)$ because the function under the integral sign is odd with respect to x_m. For $\widehat{u}$, $\widehat{f}$, $\widehat{g}$, and h_1 the integral over $B_R(0)$ in (9.8) is equal to the integral of u, f, g, h over $D = B_R(0) \cap \mathbb{R}^m_+$. Under the assumptions of the lemma, this integral vanishes. Using the inequality for $\widehat{u}$ in $D = B_R(0)$ (cf. Lemma 9.1), we obtain (9.9) for u in $D = B_R(0) \cap \mathbb{R}^m_+$. □

LEMMA 9.3. *Let $D = B_R(0)$ or $D = B_R(0) \cap \mathbb{R}^m_+$, let the functions u, f, g belong to $W^1_q(D)$, $L_q(D, \mathbb{R}^m)$, $L_q(D)$, and let their supports belong to $B_\rho(0) \cap D$, $0 < \rho < R$. Suppose that a symmetric matrix $Q(x)$ with entries $Q_{ij} \in C(\overline{D})$ is positive definite at a point $x = 0$, and that for all infinitely differentiable functions h in $\overline{D}$ vanishing near the spherical part of ∂D, the following equality holds:*

$$\int_D \left(Q_{ij} u_{x_i} h_{x_j} + f_i h_{x_i} + gh\right) dx = 0. \tag{9.10}$$

Then (9.9) also holds for all $q \in (1, \infty)$ and sufficiently small $\rho = \rho(q)$.

PROOF. We make the change of variables $x \to y = Ux$ in $\mathbb{R}^m$, where U denotes a rotation matrix such that $UQ(0)U^*$ is a diagonal matrix. Then the domain $D \equiv D_x$ becomes the domain D_y, which keeps the form of a sphere or a half-sphere. Generally speaking, the half-sphere is not determined by the relation $B_R(0) \cap \mathbb{R}^m_+$. We make the change of variables $y \to z = Vy$ in $\mathbb{R}^m$, where V is a dilation matrix along the coordinate axes such that $VUQ(0)U^*V^*$ is the identity matrix. Under such a transformation, the domain D_y goes to a domain D_z of some other geometrical form. However, for ρ small enough, it is possible to integrate (9.10) over the domain D_z which has the form of a ball $B_R(0)$ or a half-ball of the same radius. Finally, we make the change of variables $z \to x = Wz$, where $W = 1$ if $D_z = B_R(0)$ and if D_z is a half-ball, W is a rotation matrix taking this half-ball to the standard half-ball $B_R(0) \cap \mathbb{R}^m_+$. After all these operations, the equality (9.10) takes the form

$$\int_D \left(P_{ij} u_{x_i} h_{x_j} + F_i h_{x_i} + gh\right) dx = 0, \tag{9.11}$$

where D is either $B_R(0)$ or $B_R(0)\cap\mathbb{R}^m_+$. The matrix $P(x)=WVUQ(x)U^*VW^*$ is symmetric, and $P_{ij}\in C(\overline{D})$, $P(0)=1$. The functions F_i, $i=1,\dots,m$, are linear combinations with constant coefficients of the functions f_j computed in the new coordinates. The supports of the functions u, F, and g belong to $B_{\rho'}(0)\cap D$, and $\rho'=\rho'(\rho)\to 0$ as $\rho\to 0$. Therefore, for sufficiently small ρ, the integral in (9.11) is taken, in fact, over $B_{\rho'}(0)\cap D$ for sufficiently small ρ' and $x\in|P(x)-1|\le\mu(\rho)\to 0$ as $\rho\to 0$. This allows us to rewrite (9.11) in the form

$$\int_D [(\nabla u+(F+(P-1)\nabla u),\nabla h)_m+gh]\,dx=0$$

and use the result of a previous lemma to obtain

$$\|\nabla u\|_q\le C_q\left[\|F\|_q+\mu(\rho)\|\nabla u\|_q+\|g\|_q\right].$$

Using the estimate $\|F\|_q\le C(q)\|f\|_q$, taking a sufficiently small number $\rho=\rho(q)$, and returning to the old coordinates, we obtain (9.9). □

LEMMA 9.4. *Let $\omega\subset\mathbb{R}^m$ be a bounded domain with boundary $\partial\omega\in C^1$, and let a matrix $Q(x)$ with continuous entries be positive definite in $\overline{\omega}$. Suppose that functions $u\in W^1_q(\omega)$, $f\in L_q(\omega,\mathbb{R}^m)$, and $g\in L_q(\omega)$ have supports in $B_\rho(x_0)\cap\omega$, where $x_0\in\partial\omega$, and satisfy the equality*

$$\int_\omega \left(Q_{ij}u_{x_i}h_{x_j}+f^ih_{x_i}+gh\right)dx=0 \tag{9.12}$$

for any function $h\in C^1(\overline{\omega})$. Then the estimate (9.9) holds for any $q\in(1,\infty)$ and sufficiently small $\rho=\rho(q)$ in ω.

PROOF. Since $\partial\omega\in C^1$, for ρ_0 small enough there exists a $C^1(\overline{B}_{\rho_0}(x_0),\mathbb{R}^m)$-diffeomorphism $y=\psi(x)$ of the ball $B_{\rho_0}(x_0)$ such that $\psi(B_{\rho_0}(x_0)\cap\omega)\subset\mathbb{R}^m_+$ and $\psi(B_\rho(x_0)\cap\partial\omega)\subset\partial\mathbb{R}^m_+$. Since the diffeomorphism is of class C^1, the coefficients $a_{ij}(y)$ in the equality $\partial/\partial x_j=a_{ij}(y)\partial/\partial y_j$ are continuous. Assuming that $\rho<\rho_0$, we make the change of variables $x\to y=\psi(x)$ in (9.12). In the new coordinates, (9.12) can be written as follows:

$$\int_D \left(Q_{ij}a_{ik}a_{jl}u_{y_k}h_{y_l}+a_{ik}f^ih_{y_k}+gh\right)\det\dot{x}\,dy=0. \tag{9.13}$$

All the functions in (9.13) are considered in the new coordinates y, $D=B_R(0)\cap\mathbb{R}^m_+$ for some R, and the supports of the functions u, f, and g are contained in $B_{\rho'}(0)\cap D$, where $\rho'=\rho'(\rho)\to 0$ as $\rho\to 0$. Since the Jacobi matrix of the diffeomorphism ψ is nonsingular, multiplying (9.13) (if necessary) by -1, we can achieve the positive definiteness of the matrix $P_{kl}(y)=\det\dot{x}(y)Q_{ij}(x(y))a_{ik}(y)a_{jl}(y)$ at the point $y=0$. To complete the proof, it remains to apply Lemma 9.3 and return to the original coordinates. □

LEMMA 9.5. *Let $\omega \subset \mathbb{R}^m$ be a bounded domain with boundary $\partial\omega \in C^2$, let $Q(x)$ be a twice continuously differentiable positive definite matrix-valued function in $\overline{\omega}$, and let $b \in L_\infty(\omega)$, $c \in C^1(\partial\omega)$, $f \in L_q(\omega, \mathbb{R}^m)$, $g \in L_q(\omega)$. Suppose that a function $u \in W_q^1(\omega)$ satisfies the equality*

$$\int_\omega \left(Q_{ij} u_{x_i} h_{x_j} + buh + f^i h_{x_i} + gh\right) dx + \int_{\partial\omega} cuh\, dS = 0 \tag{9.14}$$

for any $h \in C^1(\overline{\omega})$. Then the estimate

$$\|\nabla u\|_q \le C_q \left(\|f\|_q + \|g\|_q + \|u\|_q\right) \tag{9.15}$$

holds for $q \in (1, \infty)$.

PROOF. Let $x_l \in \overline{\omega}$, $l = 1, \dots, N$, be a set of points such that the balls $B_\delta(x_l)$ cover $\overline{\omega}$. We divide the set of x_l into two groups y_l and z_l according to the following rule: $\overline{B}_\delta(y_l) \subset \omega$ and $\partial B_\delta(z_l) \cap \partial\omega \ne \varnothing$. Take a sufficiently small number δ and include the balls of both groups into larger balls:

$$\overline{B}_\delta(y_l) \subset B_\rho(y_l) \subset \omega, \quad \overline{B}_\delta(z_l) \subset B_\rho(z_l'), \quad z_l' \in \partial\omega,$$

where ρ is, in general, different for different balls and satisfies the assumptions of Lemma 9.3 for the first group and the assumptions of Lemma 9.4 for the second group. We assume that the functions $\varphi_l \in C^2(\overline{B}_\rho(y_l))$ have compact support and are equal to 1 in $B_\delta(y_l)$ for the first group, and the $\varphi_l \in C^2(\overline{B}_\rho(z_l'))$ have compact support and satisfy the conditions

$$\varphi_l(x) \ge 1, \quad x \in B_\delta(z_l), \qquad Q_{ij}\varphi_{lx_j}\nu_j + c\varphi_l = 0, \quad x \in \partial\omega, \tag{9.16}$$

for the second group. The construction of such functions in the case $\partial\omega \in C^2$ can be found in the Appendix. Since the further computations are similar for both groups, we will omit the subscript l in the notation of φ. We also omit the notation y_l or z_l' of the center of the ball B_ρ. In (9.14), we change h to φh. Using the formula

$$Q_{ij} u_{x_i} (\varphi h)_{x_j} = Q_{ij} (\varphi u)_{x_i} h_{x_j} - Q_{ij} \varphi_{x_i} u h_{x_j} + \left(Q_{ij} u \varphi_{x_j}\right)_{x_i} h - \left(Q_{ij}\varphi_{x_j}\right)_{x_i} uh,$$

we rewrite (9.14) in the form

$$\begin{aligned} &\int_\omega \Big[Q_{ij}(\varphi u)_{x_i} h_{x_j} + \left(f^j\varphi - 2Q_{ij}\varphi_{x_i} u\right) h_{x_j} \\ &\qquad + \Big(bu\varphi + f^j\varphi_{x_j} + g\varphi - \left(Q_{ij}\varphi_{x_j}\right)_{x_i} u\Big) h\Big]\, dx \\ &+ \int_{\partial\omega} \left(Q_{ij}\varphi_{x_j}\nu_i + c\varphi\right) uh\, dS = 0. \end{aligned} \tag{9.17}$$

For balls of the first group we have $\partial\omega \cap B_\rho = \varnothing$; hence in (9.17) only the integral over ω remains. We can assume that the integral is taken over $D = B_R(0)$. Using Lemma 9.3, in the ball B_ρ we obtain the estimate

$$\|\nabla(\varphi u)\|_q \le C_q \left(\|f\|_q + \|g\|_q + \|u\|_q\right), \tag{9.18}$$

which yields

$$\|\nabla u\|_q(B_\delta) \le C_q \left[\|f\|_q(\omega) + \|g\|_q(\omega) + \|u\|_q\right]. \tag{9.19}$$

For the balls of the second group, the integral over $\partial\omega$ vanishes in view of the boundary condition in (9.16). Applying Lemma 9.4, we arrive at the estimate (9.18), which implies (9.19) with $B_\delta \cap \omega$ instead of B_δ on the left-hand side. From these estimates and the triangle inequality

$$\|\nabla u\|_q(\omega) \leq \left\| \sum_l \chi_l(x)|\nabla u| \right\|_q (\omega) \leq \sum_l \|\nabla u\|_q \left(B_\delta(x_l) \cap \omega\right),$$

where $\chi_l(x)$ denotes the characteristic function of the set $B_\delta(x_l) \cap \omega$, we obtain (9.15). □

Now we are in a position to complete the proof of (9.6). Since Lemma 9.5 implies the inequality

$$\|p\|_{1,q} \leq C_q \left(\|f\|_q + \|p\|_q\right), \tag{9.20}$$

it remains to show that for the solution $p(x)$ to the problem (9.4) from Theorem 9.1, the second term on the right-hand side of (9.20) may be omitted.

Assume, on the contrary, that (9.6) with the above term eliminated is false. There exists a sequence f_j of the right-hand sides of the problem (9.4) such that the constructed solutions p_j satisfy the estimate $\|p_j\|_{1,q} > j\|f_j\|_q$ for $j = 1, \dots$ and some q. In this case, the functions $p_j = p_j\|p_j\|_{1,q}^{-1}$ are solutions to (9.4) with right-hand sides $\widehat{f}_j = f_j\|p_j\|_{1,q}^{-1}$ and $1 = \|\widehat{p}_j\|_{1,q} > j\|\widehat{f}_j\|_q$. Thus, $\|\widehat{f}_j\|_q \to 0$ as $j \to \infty$ and $\|\widehat{p}_j\|_{1,q} = 1$. Since $q > 1$, there exists a subsequence (we preserve the same notation) $\widehat{p}_j(x)$ that converges weakly to a function $\widehat{p} \in W_q^1(\omega)$ in $W_q^1(\omega)$. Since the embedding of $W_q^1(\omega)$ into $L_q(\omega)$ is compact, the inequality (9.20) written for the solution $\widehat{p}_j - \widehat{p}_l$ corresponding to the right-hand side $\widehat{f}_j - \widehat{f}_l$ implies that the sequence of solutions $\widehat{p}_j$ is a Cauchy sequence in the space $W_q^1(\omega)$. Therefore, $\widehat{p}_j \to \widehat{p}$ in $W_q^1(\omega)$, and $\|\widehat{p}\|_{1,q} = 1$. Passing to the limit in (9.7) written for $\widehat{p}_j$ and $\widehat{f}_j$, we conclude that $\widehat{p}$ is a nonzero solution to the homogeneous problem (9.4). We note that $\widehat{p}(x)$, being a solution to some elliptic problem, belongs to $C^{k+1,\varepsilon}(\overline{\omega})$. A contradiction arises since the homogeneous problem (9.4) has the zero solution either automatically or under the additional assumption (9.5), which holds for $\widehat{p}(x)$ provided that it holds for all $\widehat{p}_j(x)$.

§10. Weyl decomposition of L and stability

In §3, we considered the Weyl decomposition, which presents a general form of a function $f \in L_2(\omega, \mathbb{R}^m)$ orthogonal to the set of all smooth solenoidal compactly supported vector-valued functions in $L_2(\omega, \mathbb{R}^m)$. The goal of this section is to describe functions $f \in L_2(\omega, \mathbb{R}^n)$ that are orthogonal to the space $J^{k,\varepsilon}(\partial\omega, \varnothing)$ in $L_2(\omega, \mathbb{R}^n)$. The description will be obtained as a consequence of a more general result concerning the theorem on stability of the Weyl decomposition of the operator L in the space $L_q(\omega, \mathbb{R}^n)$, $1 < q < \infty$. The proof of the theorem is divided into several steps.

LEMMA 10.1. *Let $u \in C^{k,\varepsilon}(\overline{\omega}, \mathbb{R}^n)$ be such that $Lu = 0$ and $(A\nu, u)_n\big|_{\partial\omega} = 0$. Then there exists a sequence $v_j \in J^{k,\varepsilon}(\partial\omega, \varnothing)$, $j = 1, \dots, n$, such that $\|u - u_j\|_q \to 0$ as $j \to \infty$ for every $q \in [1, \infty)$.*

PROOF. Let $\varphi \in C^\infty(\mathbb{R}^1_+)$ and let $\varphi(t) = t$ for $t \le 1/2$, $\varphi(t) \equiv 1$ for $t \ge 1$, and $0 < \varphi(t) \le 1$ for $t > 0$. For sufficiently small δ we define the function $\chi_\delta \in C^{k+1,\varepsilon}(\overline{\omega})$ as follows: $\chi_\delta(x) = \varphi(\delta^{-1}\rho(x))$, where $\rho(x)$ is the distance from $x \in \overline{\omega}$ to $\partial\omega$ (cf. the Appendix). It is obvious that $\chi_\delta(x)$ differs from 1 only in some boundary strip Γ_δ of width δ. Let $\widehat{u}(x) = \chi_\delta(x)u(x)$. Let $\widehat{v}(x)$ be the solution to the problem

$$L\widehat{v} = L\widehat{u}, \quad \widehat{v}\big|_{\partial\omega} = 0, \quad \widehat{v} \in C^{k,\varepsilon}(\overline{\omega}, \mathbb{R}^n),$$

from Theorem 5.1, and let $v(x)$ be defined by the equality $v(x) = \widehat{u}(x) - \widehat{v}(x)$. Since $L\widehat{u} = (u, A\nabla\chi_\delta)_n = \delta^{-1}\dot{\varphi}(\delta^{-1}\rho)(u, A\nabla\rho)_n$, we see that $L\widehat{u} \in C^{k\varepsilon}(\overline{\omega})$ differs from zero only in Γ_δ and satisfies the estimate

$$|L\widehat{u}(x)| \le C\delta^{-1}|(u(x), A(x)\nabla\rho(x))_n|, \qquad C \ne C(\delta),$$

there. The function $(u, A\nabla\rho)_n$ belongs to $C^1(\Gamma_\delta)$. Since $\nabla\rho(x) = -\nu(x)$ on $\partial\omega$, we have $(u, A\nabla\rho)_n\big|_{\partial\omega} = 0$. Therefore,

$$\big|(u(x), A(x)\nabla\rho(x))_n\big| \le C\,\mathrm{dist}(x, \partial\omega) \le C\delta \quad \text{for} \quad x \in \Gamma_\delta.$$

The last two inequalities yield the estimate $\|L\widehat{u}\|_r \le C_r\delta^{1/r}$, $r \in (1, \infty)$, which leads to the estimate for the solution $\widehat{v}$: $\|\widehat{v}\|_{1,r} \le C_r\delta^{1/r}$ with some positive constant $C_r \ne C_r(\delta)$. Using the triangle inequality, we get

$$\|u - v\|_q \le \|u - \widehat{u}\|_q + \|\widehat{v}\|_q \le 2\|u\|_q(\Gamma_\delta) + \|\widehat{v}\|_q.$$

If r is such that the embedding operator of $W^1_r(\omega)$ into $L_q(\omega)$ is bounded, then

$$\|u - v\|_q \le 2\|u\|_q(\Gamma_\delta) + C_{r,q}\delta^{1/r}, \quad C_{r,q} \ne C_{r,q}(\delta).$$

The right-hand side of this inequality converges to zero as $\delta \to 0$. Therefore, for $v_j(x)$ we can take $v(x)$ with $\delta = \delta_j \to 0$ as $j \to \infty$. □

LEMMA 10.2. *Let a function $u \in C^{k,\varepsilon}(\overline{\omega}, \mathbb{R}^n)$ satisfy the conditions*

$$Lu = 0, \quad (u, A\nu)_n\big|_{\partial\omega} = 0, \quad \Big|\int_\omega (u, v)_n\,dx\Big| \le \alpha\|v\|_{q'} \tag{10.1}$$

for all $v \in J^{k,\varepsilon}(\partial\omega, \varnothing)$, where the index q' is conjugate to $q \in (1, \infty)$. Then $u(x)$ satisfies the estimate $\|u\|_q \le C_q\alpha$, $C_2 = 1$.

PROOF. By Lemma 10.1, we can assume that the inequality

$$\Big|\int_\omega (u, v)_n\,dx\Big| \le \alpha\|v\|_{q'} \tag{10.2}$$

holds for all functions $v \in C^{k,\varepsilon}(\overline{\omega}, \mathbb{R}^n)$ such that

$$Lv = 0, \qquad (A\nu, v)_n\big|_{\partial\omega} = 0. \tag{10.3}$$

To prove the lemma in the case $q = 2$, it suffices to substitute $v = u$ in (10.2). We consider the case $q \ne 2$. Let $f \in C_0^\infty(\omega, \mathbb{R}^n)$ and let $p(x)$ be a solution to the problem (9.1) about the decomposition of f into the sum $v + L^*p$, where v satisfies (10.3). Substituting $v = f - L^*p$ in (10.2), taking into account the fact that the integral of $(L^*p, u)_n$ over ω vanishes, and recalling the last estimate for $\|L^*p\|_{q'}$ in (9.2), we conclude that

$$\left|\int_\omega (u, f)_n \, dx\right| \leq \alpha \|f - L^*p\|_{q'} \leq \alpha(1 + C'_{q'})\|f\|_{q'} \equiv \alpha C_q \|f\|_{q'}.$$

We extend the last inequality by continuity to the set of all functions $f \in L_{q'}(\omega, \mathbb{R}^n)$ and put

$$f(x) = \begin{cases} 0 & \text{for } u(x) = 0, \\ |u(x)|^{q-2}u(x) & \text{for } u(x) \neq 0. \end{cases}$$

The last inequality is the required estimate for u. □

THEOREM 10.1. *Let $f \in L_q(\omega, \mathbb{R}^n)$, $q \in (1, \infty)$, satisfy the inequality*

$$\left|\int_\omega (f, \nu)_n \, dx\right| \leq \alpha \|v\|_q$$

for all $v \in J^{k,\varepsilon}(\partial\omega, \varnothing)$. Then there is a function $p \in W^1_q(\omega)$ with $\|f - L^*p\|_q \leq \alpha C_q$, $C_2 = 1$.

PROOF. For each $\varkappa > 0$ we can find a function $f_\varkappa \in C_0^\infty(\omega, \mathbb{R}^n)$ such that $\|f - f_\varkappa\|_q < \varkappa$ and, with the help of (9.1), represent $f_\varkappa(x)$ in the form

$$\begin{aligned} &f_\varkappa = u_\varkappa + L^*p_\varkappa, \quad u_\varkappa \in C^{k,\varepsilon}(\overline{\omega}, \mathbb{R}^n), \quad p_\varkappa \in C^{k+1,\varepsilon}(\overline{\omega}), \\ &Lu_\varkappa = 0, \quad (u_\varkappa, A\nu)_n\big|_{\partial\omega} = 0. \end{aligned} \tag{10.4}$$

Then $f(x)$ satisfies the equality

$$f = u_\varkappa + L^*p_\varkappa + (f - f_\varkappa). \tag{10.5}$$

Multiplying this equality by v and integrating over ω, we obtain

$$\left|\int_\omega (u_\varkappa, v)_n \, dx\right| \leq (\alpha + \varkappa)\|v\|_{q'}.$$

Lemma 10.2 implies the estimate $\|u_\varkappa\|_q \leq C_q(\alpha + \varkappa)$, which, together with (10.5), yields the relation

$$\|f - L^*p_\varkappa\|_q \leq \varkappa + C_q(\alpha + \varkappa). \tag{10.6}$$

Take two values of $\varkappa$: $\varkappa = \varkappa_1$ and $\varkappa = \varkappa_2$. From (10.4) it follows that

$$u_{\varkappa_1} - u_{\varkappa_2} = (f_{\varkappa_1} - f_{\varkappa_2}) - L^*(p_{\varkappa_1} - p_{\varkappa_2}). \tag{10.7}$$

Multiplying the last equality by v and integrating over ω, we derive the estimate

$$\left|\int_\omega (u_{\varkappa_1} - u_{\varkappa_2}, v)_n \, dx\right| \leq \|f_{\varkappa_1} - f_{\varkappa_2}\|_q \|v\|_{q'}.$$

By Lemma 10.2, we obtain

$$\|u_{\varkappa_1} - u_{\varkappa_2}\|_q \leq C_q \|f_{\varkappa_1} - f_{\varkappa_2}\|_q. \tag{10.8}$$

Rewriting (10.7) in the form

$$L^*p_{\varkappa_1} - L^*p_{\varkappa_2} = (f_{\varkappa_1} - f_{\varkappa_2}) - (u_{\varkappa_1} - u_{\varkappa_2})$$

and using (10.8), we find that

$$\|L^* p_{\varkappa_1} - L^* p_{\varkappa_2}\|_q \leq (1 + C_q) \|f_{\varkappa_1} - f_{\varkappa_2}\|_q. \tag{10.9}$$

Let $\varkappa_j \to 0$. Then $f_{\varkappa_j}$ is a Cauchy sequence in L_q and, by (10.9), $L^* p_{\varkappa_j}$ is a Cauchy sequence in L_q. If the problem $L^* p = 0$, $p \in C^1(\overline{\omega})$, has only the zero solution, then $p_{\varkappa_j}$ is a Cauchy sequence in $W_q^1(\omega)$ by Lemma 2.3. If the problem has a nonzero solution, then $p_{\varkappa_j}$ is a Cauchy sequence by Lemma 2.4 and the condition (9.5). We see that in both cases, $p_{\varkappa_j} \to p$ in $W_q^1(\omega)$ as $\varkappa_j \to 0$. We write (10.6) for $\varkappa = \varkappa_j$. Passing to the limit in this inequality as $\varkappa_j \to 0$, we complete the proof of the theorem. □

A special case of this theorem ($\alpha = 0$) is similar to the Weyl decomposition of the operator L. The case $\alpha \neq 0$ describes the stability of this decomposition.

CHAPTER 2

Nonlinear Perturbations of the Operator div

Introduction

In this chapter, we consider those small nonlinear perturbations of the operator L that contain the first order derivatives.

The main purpose is to describe the set of all solutions to the perturbated problem that lie in a sufficiently small neighborhood of a fixed solution z. Of special interest are two extreme cases. In the first case, the set of all solutions looks like a piece of a smooth surface lying in some function space and containing z. In the second case, the solution z is isolated.

A natural tool for studying such problems is the implicit function theorem and the Lyapunov–Schmidt procedure. In §1, we give all necessary information on the differential calculus in a normed space. In §2, we study the differentiability of mappings to be used below. In §3, we prove the implicit function theorem. In §§4 and 5, we study the set of all solutions to nonlinear equations in the scale of Banach spaces under two different assumptions on properties of the linearized problem. In §6, the bifurcation problem is considered. This auxiliary material is used in §7 for a local description of the set of all solutions to a nonlinear first-order differential equation whose linearization is given by the operator L. In §8, we introduce the notion of a rigid condition on the set of solutions to a nonlinear problem, and study the rigidity of various conditions. In §9, we consider specific nonlinear problems, including invariants of the metric tensor of the required mapping $y(x)$ from a domain ω to the space $\mathbb{R}^m$. In §10, the above nonlinear equations appear as constraints in variational problems for integral functionals. Necessary and sufficient conditions for an extremum are presented, and the method of Lagrange multipliers is discussed. As an application, we consider variational problems in the theory of the elasticity of an incompressible medium.

§1. Foundation of differential calculus in a normed space

1.1. The differential of a mapping. We consider normed spaces X, Y and an open set $U \subset X$. A mapping $F : U \to Y$ is called differentiable at a point $x \in U$ if there exists a bounded linear operator $L_x : X \to Y$ such that

$$F[x+h] - F[x] - L_x h = \alpha(x,h), \quad x+h \in U,$$
$$\|\alpha(x,h)\| \, \|h\|^{-1} \to 0 \text{ as } h \to 0.$$

The expression $L_x h$ is called the differential of F at the point x, and the linear operator L_x itself is called the derivative and is denoted by $F'[x]$.

If a mapping F is differentiable at a point x, its derivative is uniquely determined. Indeed, let $F[x+h]-F[x]=L_x^{(1)}h+\alpha_1(x,h)=L_x^{(2)}h+\alpha_2(x,h)$. Then $(L_x^{(1)}-L_x^{(2)})h=\alpha(x,h)$ and $\|\alpha(x,h)\|\,\|h\|^{-1}\to 0$ as $h\to 0$. We take $h=te$, where $e\in X$ is an arbitrary vector and t is a sufficiently small parameter. Then $(L_x^{(1)}-L_x^{(2)})e=\alpha(x,te)t^{-1}$ as $t\to 0$. This means that $L_x^{(1)}$ and $L_x^{(2)}$ coincide.

A mapping $F[x]$ is said to be continuously differentiable in U if $F'[x]\in L(X,Y)$ depends continuously on x in the operator norm.

In the case of a one-dimensional space X, any linear operator $L:X\to Y$ can be defined by the equality $Lh=yh$, $h\in X=\mathbb{R}$, $y\in Y$. Therefore, the operator L can be identified with the vector y. This gives us the possibility of introducing an equivalent definition of the derivative of F in the case of a one-dimensional space X, i.e., $F'[x]=\lim h^{-1}(F[x+h]-F[x])$ as $h\to 0$.

1.2. The derivative of a composite function. We consider normed spaces X, Y, Z, open sets $U\subset X$, $V\subset Y$, and two mappings $F:U\to Y$, $G:V\to Z$. Let $F[x_0]=y_0\in V$ at $x_0\in U$ and let the mappings F and G be differentiable at x_0 and y_0 respectively. Then the mapping $H=G\circ F$ is defined in a neighborhood of x_0, is differentiable there, and satisfies the relation $H'[x_0]=G'[y_0]F'[x_0]$ at x_0. Indeed, since F is differentiable at x_0, it is continuous at x_0. Therefore, the smallness of $\|x-x_0\|$ implies the smallness of $\|F[x]-y_0\|$. Thus, under the mapping F, a sufficiently small neighborhood of x_0 goes to V, which allows us to define the composition $H=G\circ F$ in this neighborhood. Further, we have

$$\begin{aligned}H[x_0+h]-H[x_0]&=G\big[F[x_0+h]\big]-G\big[F[x_0]\big]\\&=G\big[y_0+F'[x_0]h+\alpha_F(x_0,h)\big]-G[y_0]\\&=G'[y_0]F'[x_0]h+G'[y_0]\alpha_F(x_0,h)\\&\quad+\alpha_G\big(y_0,F'[x_0]h+\alpha_F(x_0,h)\big)\\&\equiv G'[y_0]F'[x_0]h+\alpha_H(x_0,h).\end{aligned}$$

The subscript to α indicates the mapping to which it refers.

1.3. Law of the mean. Let $U\subset X$ be an open set. We assume that points x_1, x_2 and the segment $[x_1,x_2]=\{x_t=(1-t)x_1+tx_2 : 0\leqslant t\leqslant 1\}$ joining them lie in U. We also assume that a mapping $F:U\to Y$ is differentiable at every point $x_t\in[x_1,x_2]$. Then

$$\big\|F[x_2]-F[x_1]\big\|\le\sup_{0\le t\le 1}\big\|F'[x_t]\big\|\,\|x_1-x_2\|. \tag{1.1}$$

Indeed, for any bounded linear functional l the function $\varphi(t)=l(F[x_t])$ is continuous in $[0,1]$ and has finite derivative at every point of $[0,1]$. Therefore, there exists $t^*\in[0,1]$ such that $\varphi(1)-\varphi(0)=\varphi'(t^*)$. The last equality implies

$$\big|l\big(F[x_2]-F[x_1]\big)\big|\le\|l\|\sup_{0\le t\le 1}\big\|F'[x_t]\big\|\,\|x_2-x_1\|.$$

Taking for l a nonzero functional such that

$$\big|l\big(F[x_2]-F[x_1]\big)\big|=\|l\|\,\big\|F[x_2]-F[x_1]\big\|,$$

we arrive at the required inequality. Applying our estimate to the mapping $x \to F[x] - Tx$, $T \in L(X,Y)$, we get

$$\big\|F[x_2] - F[x_1] - T(x_2 - x_1)\big\| \le \sup_{0\le t\le 1} \big\|F'[x_t] - T\big\| \, \|x_2 - x_1\|. \tag{1.2}$$

1.4. The higher order derivatives. We consider an open set $U \subset X$ and a mapping $F : U \to Y$ differentiable at each point $x \in U$. The derivative $F'[x]$ belongs to $L(X,Y)$ at every x. Hence $F'[\cdot]$ can be regarded as a mapping from the set U to the space $L(X,Y)$, and the question of its differentiability arises. The derivative (if it exists) $F''[x_0]$ belongs to $L(X, L(X,Y))$ at x_0 and depends on x_0. We give a more convenient interpretation of this space, representing it as a space of bilinear mappings from X into Y.

A mapping $B(\cdot,\cdot) : X \times X \to Y$ is called bilinear if it is a linear function of each of its arguments and satisfies the condition

$$\|B\| = \sup_{x_1\ne 0,\, x_2\ne 0} \|B(x_1,x_2)\| \, \|x_1\|^{-1}\|x_2\|^{-1} < \infty.$$

Bilinear mappings form the normed linear space $L(X^2, Y)$ equipped with the norm $\|\cdot\|$ from the last relation. For each $A \in L(X, L(X,Y))$ and $x_1 \in X$ the expression Ax_1 belongs to $L(X,Y)$ and depends linearly on x_1. Therefore, the mapping

$$B(x_1,x_2) \equiv (Ax_1)x_2 \tag{1.3}$$

takes its values in Y and is a linear function of x_1 and of x_2. Further,

$$\big\|B(x_1,x_2)\big\|_Y \le \big\|Ax_1\big\|_{L(X,Y)} \big\|x_2\big\|_X \le \|A\|_{L(X,L(X,Y))} \big\|x_1\big\|_X \big\|x_2\big\|_X,$$

where the subscript on the norm sign indicates the corresponding space. Consequently, $B(x_1,x_2)$ is a bilinear mapping. Arguing as above, with every element $A \in L(X, L(X,Y))$ we associate a bilinear form $B \in L(X^2,Y)$ such that $\|B\|_{L(X^2,Y)} \le \|A\|_{L(X,L(X,Y))}$. This correspondence is linear.

The converse assertion is also true. Let $B \in L(X^2,Y)$. We fix a point $x_1 \in X$. Then $(Ax_1)(\cdot) \equiv B(x_1,\cdot)$ (i.e., formula (1.3) written from back to front) belongs to $L(X,Y)$ and depends linearly on x_1. Moreover,

$$\big\|(Ax_1)x_2\big\|_Y = \big\|B(x_1,x_2)\big\|_Y \le \big\|B\big\|_{L(X^2,Y)} \big\|x_1\big\|_X \big\|x_2\big\|_X.$$

Hence

$$\big\|Ax_1\big\|_{L(X,Y)} \le \big\|B\big\|_{L(X^2,Y)} \big\|x_1\big\|_X, \quad \big\|A\big\|_{L(X,L(X,Y))} \le \big\|B\big\|_{L(X^2,Y)}.$$

Thus, for every $B \in L(X^2,Y)$ formula (1.3) defines an operator $A \in L(X, L(X,Y))$. The norm of A is majorized by the norm of B. This shows that (1.3) is an isomorphism between $L(X^2,Y)$ and $L(X, L(X,Y))$.

The notion of the nth derivative of a mapping $F : U \to Y$ can be introduced in a similar way. If the nth derivative of F exists at $x_0 \in U$, then $F^{(n)}[x_0] \in L\big(X, L(X, \ldots, L(L,Y))\big) \equiv L_n(X,Y)$, where the symbol X appears n times in the expression $\big(X, L(X, \ldots, L(L,Y))\big)$. To interpret the space $L_n(X,Y)$, we introduce the notion of an n-linear mapping from X into Y. The mapping $B(x_1,\ldots,x_n)$ of $x_j \in X$, $j = 1,\ldots,n$, is called n-linear if it is a linear function of each of its arguments and satisfies the condition

$$\|B\| = \sup_{x_j\ne 0,\, j=1,\ldots,n} \|B(x_1,\ldots,x_n)\| \, \|x_1\|^{-1} \cdots \|x_n\|^{-1} < \infty.$$

The set of all n-linear mappings is the linear space $L(X^n, Y)$ equipped with the norm from the last relation. Proceeding by induction, we can prove that the space $L_n(X,Y)$ is isometrically isomorphic to $L(X^n, Y)$. Indeed, for $n = 2$ and any X, Y the assertion is true. We assume that it is true for $n = k$, i.e., the spaces

$$L(X, L(X, \dots, L(X, Z))), \quad L(X^k, Z) \tag{1.4}$$

are isometrically isomorphic for any X and Z, where the symbol X appears k times in the expression $L(X, L(X, \dots, L(X, Z)))$. The relations

$$\begin{gathered} B_{k+1}(x_1, \dots, x_{k+1}) = B_k(x_1, \dots, x_k)x_{k+1}, \\ B_{k+1} \in L(X^{k+1}, Y), \quad B_k \in L(X^k, L(X, Y)), \end{gathered} \tag{1.5}$$

which are similar to (1.3), allow us, arguing as above, to prove that the spaces

$$L(X^{k+1}, Y), \quad L(X^k, L(X, Y)) \tag{1.6}$$

are isometrically isomorphic. Substituting $Z = L(X, Y)$ in (1.4) and taking (1.6) into account, we obtain (1.4) with k replaced by $k + 1$.

The mapping $F : U \to X$ is said to be k times continuously differentiable in U if the kth order derivative $F^{(k)}[x]$ exists at each point $x \in U$ and depends continuously on x. We prove that for a k-times continuously differentiable mapping F the expression $F^{(k)}[x](x_1, \dots, x_k)$ is symmetric with respect to x_i, $i = 1, \dots, k$. Let $\Delta_y f[x] = f[x + y] - f[x]$ and let $L(y_1, \dots, y_k)$ be a k-linear mapping. Using (1.2) with $x_2 = y_k + x$, $x_1 = x$, $T(\cdot) = (y_1, \dots, y_{k-1}, \cdot)$ and taking $\Delta_{y_{k-1}} \dots \Delta_{y_1} F[x]$ for $F[x]$, we get

$$\begin{aligned} &\|\Delta_{y_k}\Delta_{y_{k-1}} \dots \Delta_{y_1} F[x] - L(y_1, \dots, y_k)\| \\ &\qquad \le \sup_{0 \le t \le 1} \|\Delta_{y_{k-1}} \dots \Delta_{y_1} F'[x + t y_k] y_k - L(y_1, \dots, y_k)\|. \end{aligned}$$

Iterating the last inequality, we arrive at the relation

$$\begin{aligned} &\|\Delta_{y_k}\Delta_{y_{k-1}} \dots \Delta_{y_1} F[x] - L(y_1, \dots, y_k)\| \\ &\qquad \le \sup_{0 \le t_1, \dots, t_k \le 1} \Big\|F^{(k)}\Big[x + \sum_{j=1}^{k} t_j y_j\Big](y_1, \dots, y_k) - L(y_1, \dots, y_k)\Big\|. \end{aligned}$$

Setting $L(y_1, \dots, y_k) = F^{(k)}[x](y_1, \dots, y_k)$, we obtain

$$\|\Delta_{y_k} \dots \Delta_{y_1} F[x] - F^{(k)}[x](y_1, \dots, y_k)\| = o(1)\|y_1\| \dots \|y_k\| \tag{1.7}$$

as $\sum_{y=1}^{k} \|y_j\| \to 0$. The relation (1.7) uniquely determines $F^{(k)}[x](y_1, \dots, y_k)$. Indeed, let a k-linear mapping $B(y_1, \dots, y_k)$ satisfy (1.7). Then

$$\|B(y_1, \dots, y_k) - F^{(k)}(y_1, \dots, y_k)\| = o(1)\|y_1\| \dots \|y_k\|.$$

We fix z_i, $i = 1, \dots, k$, and put $y_i = t z_i$. Dividing both sides of the last equality by t^k and passing to the limit as $t \to 0$, we get $B(z_1, \dots, z_k) = F^{(k)}[x](z_1, \dots, z_k)$. Since $\Delta_{y_k} \dots \Delta_{y_1} F[x]$ remains unchanged under any permutation of y_i, the relation (1.7) is satisfied by $F^{(k)}[x](y_{i_1}, \dots, y_{i_k})$ provided that $F^{(k)}[x](y_1, \dots, y_k)$ satisfies (1.7), where $y_{i_1}, \dots, y_{i_k}$ is any permutation of $y_1, \dots, y_k$. Therefore,

$$F^{(k)}[x](y_1, \dots, y_k) = F^{(k)}[x](y_{i_1}, \dots, y_{i_k}).$$

1.5. The Taylor formula with remainder in the Peano form. Let $U \subset X$ be an open set and let a mapping $F : U \to Y$ have an $(n-1)$th derivative in U and an nth derivative at $x \in U$. Then

$$\begin{gathered} F[x+h] - F[x] - \sum_{j=1}^{n} (j!)^{-1} F^{(j)}[x](h, \dots, h) = \alpha(x,h), \\ \|\alpha(x,h)\| \, \|h\|^{-n} \to 0 \text{ as } h \to 0. \end{gathered} \tag{1.8}$$

To prove (1.8), we note that for a fixed x the function $\Phi[h] \equiv \alpha(x,h)$ is defined for all $h \in X$ of norm small enough, and has an $(n-1)$th derivative with respect to h and an nth derivative with respect to h at $h = 0$; moreover,

$$\Phi^{(j)}[0] = 0, \quad j = 0, \dots, n. \tag{1.9}$$

It remains to prove that (1.9) implies

$$\|\Phi[h]\| = o(\|h\|^n) \text{ as } h \to 0. \tag{1.10}$$

We proceed by induction. For $n = 1$ the assertion is true because of the definition of the derivative. We assume that the assertion is true for any mapping Φ satisfying (1.9) with $n = k$ and show its validity for $k+1$. Indeed, from (1.9) with $n = k+1$ it follows that the mapping $\Phi'[h]$ satisfies some condition similar to (1.8) with $n = k$. By the induction hypothesis, $\|\Phi'[h]\| = o(\|h\|^k)$. By (1.1), $\|\Phi[h]\| = o(\|h\|^{k+1})$ as $h \to 0$.

Finally, we note that (1.8) implies the estimates

$$\Big\| F[x+h] - F[x] - \sum_{j=1}^{n-1} (j!)^{-1} F^j[x](h, \dots, h) \Big\| \le C \|h\|^n \tag{1.11}$$

and that, generally speaking, the existence of the nth derivative of F at x does not follow from the decomposition

$$\begin{gathered} F[x+h] = F[x] + \sum_{j=1}^{n} (j!)^{-1} b^j[x](h, \dots, h) + \alpha(x,h), \\ b_j[x] \in L(X^j, Y), \quad \|\alpha(x,h)\| \, \|h\|^{-n} \to 0 \text{ as } h \to 0 \end{gathered} \tag{1.12}$$

even in the case $X = Y = \mathbb{R}$ (cf. [**8**]). However, if an $(n-1)$-times differentiable mapping F defined in U has an nth derivative at $x \in U$, then (1.12) holds only if $b_j = F^{(j)}[x]$.

1.6. Partial derivatives. Let X, Y, and Z be normed spaces, let $U \subset X \times Y$ be an open set, and let $F : U \to Z$ be a k-times differentiable mapping. We can define a k-linear mapping $X \times Y \to Z$ depending on the pair $\{x, y\} \in U$, i.e., a mapping $F^{(k)}[x,y](h_1, \dots, h_k)$, where $h_i = \{\xi_i, \eta_i\} \in X \times Y$. Since $F^{(k)}$ is polylinear and symmetric, we find that

$$F^{(k)}[x,y](h_1, \dots, h_k) = \sum_{j=0}^{k} \binom{k}{j} D_x^j D_y^{k-j} F[x,y](\xi_1, \dots, \xi_j, \eta_{j+1}, \dots, \eta_k),$$

where $\binom{k}{j}$ is the binomial coefficient and

$$D_x^j D_y^{k-j} F[x,y](\xi_1, \dots, \xi_j, \eta_{j+1}, \dots, \eta_k) = F^{(k)}[x,y](\xi_1, \dots, \xi_j, \eta_{j+1}, \dots, \eta_k).$$

For each pair $\{x, y\} \in U$ the expressions $D_x^j D_y^{k-j} F[x, y] \in L(X^j \times Y^{k-j}, Z)$ are called the partial derivatives of the mapping F. For the first and second order partial derivatives we also use the ordinary notation F_x, F_y and F_{xx}, F_{xy}, F_{yy}.

1.7. Independence of the derivative of the choice of norm. Let X and Y be linear spaces each of which is equipped with two norms, $\|\cdot\|_1$, $\|\cdot\|_2$ and $|\cdot|_1$, $|\cdot|_2$ respectively. Furthermore, let $|\cdot|_1 \le C|\cdot|_2$. We assume that $F : X \to Y$, being a mapping between normed spaces equipped with the norms supplied by the subscripts 1 and 2, is differentiable at a point $x_0 \in X$. We prove that the derivative $F'[x_0]$ is independent of the choice of the subscript, i.e., of the choice of the norm. Indeed, let A_1 and A_2 be the derivatives corresponding to the norms supplied with subscripts 1 and 2 respectively. Then

$$\begin{aligned} F[x_0 + h] - F[x_0] - A_1 h = \alpha_1(h), \quad & |\alpha_1(h)|_1 \|h\|_1^{-1} \to 0 \text{ as } \|h\|_1 \to 0, \\ F[x_0 + h] - F[x_0] - A_2 h = \alpha_2(h), \quad & |\alpha_2(h)|_2 \|h\|_2^{-1} \to 0 \text{ as } \|h\|_2 \to 0. \end{aligned}$$

Replacing h by th, we find that

$$\begin{aligned} |A_1 h - A_2 h|_1 &\le |t^{-1}\alpha_1(th)|_1 + |t^{-1}\alpha_2(th)|_1 \\ &\le |\alpha_1(th)|_1 \|th\|_1^{-1} \|h\|_1 + C|\alpha_2(th)|_2 \|th\|_2^{-1} \|h\|_2 \to 0 \text{ as } \|h\|_2 \to 0, \end{aligned}$$

which means that A_1 and A_2 coincide.

If F, being a mapping of X into Y with the norms supplied with by the subscripts 1 and 2 respectively, has r derivatives at x_0, then, repeating the above arguments on the Taylor expansions of power $2, \dots, r$ subsequently, we can prove that $F^{(r)}[x_0]$ is independent of the choice of the norm.

§2. Examples of differentiable mappings

2.1. Differentiable mappings of Hölder spaces. Let $\omega \subset \mathbb{R}^m$ be a bounded domain. We fix a vector-valued function $z \in C^{k,\varepsilon}(\overline{\omega}, \mathbb{R}^n)$ and introduce the open set

$$V_\rho = \{(p, q) \in \mathbb{R}^{nm} \times \mathbb{R}^n : |p - \dot{z}(x)| + |q - z(x)| < \rho\}. \tag{2.1}$$

Let F be a function of $p, q \in V_\rho$, $x \in \overline{\omega}$ such that

$$\begin{gathered} F(p, q, x) \text{ has a continuous } (r+1)\text{th order derivative for all } p, q \in \overline{V}_\rho, \\ D_p^\alpha D_q^\beta F \in C^{k-1,\alpha}(\overline{V}_\rho \times \overline{\omega}), \quad |\alpha| + |\beta| \le r + 1. \end{gathered} \tag{2.2}$$

Under these assumptions, the Taylor expansion takes place:
(2.3)

$$\begin{aligned} F(p + \delta p, q + \delta q, x) &= F(p, q, x) + \sum_{j=1}^{n} (j!)^{-1} F^{(j)}(p, q, x)(\delta p, \delta q) \\ &\quad + (n!)^{-1} \int_0^1 F^{(n+1)}(p + t\delta p, q + t\delta q, x)(\delta p, \delta q)(1 - t)^n \, dt, \end{aligned}$$

where $n = 1, \dots, r, p, q \in \overline{V}_\rho$, $p + \delta p, q + \delta q \in \overline{V}_\rho$, and $F^{(j)}(p, q, x)(\delta p, \delta q)$ denote terms containing the derivatives $D_p^\alpha D_q^\beta F(p, q, x)$, $|\alpha| + |\beta| = j$. On the open set

$$U_\rho = \{y \in C^{k,\varepsilon}(\overline{\omega}, \mathbb{R}^n) : \dot{y}, y \in V_\rho \text{ at } x \in \overline{\omega}\}, \tag{2.4}$$

we define the mapping $F : U_\rho \to C^{k-1,\varepsilon}(\overline{\omega})$ by the equality

$$F[y] = F\big(\dot y(x), y(x), x\big). \tag{2.5}$$

LEMMA 2.1. *The mapping $F[y]$ has r continuous derivatives on U_ρ, and*

$$F^{(j)}[y](h,\dots,h) = F^{(j)}\big(\dot y(x), y(x), x\big)(\dot h, h). \tag{2.6}$$

PROOF. By (2.2), we have

$$F^{(j)}\big(\dot y(x), y(x), x\big)(\cdot,\cdot) \in L\big(\big(C^{k,\varepsilon}(\overline{\omega},\mathbb{R}^n)\big)^j, C^{k-1,\varepsilon}(\overline{\omega})\big),$$
$$j = 1,\dots,r+1, \quad y \in U_\rho.$$

From (2.3) for $n = 1$ we obtain

$$F(\dot y + \dot h, y + h, x) - F(\dot y, y, x) - F^{(1)}(\dot y, y, x)(\dot h, h)$$
$$= \int_0^1 (1-t) F^{(2)}(\dot y + t\dot h, y + th, x)(\dot h, h)\, dt.$$

Taking the $C^{k-1,\varepsilon}(\overline{\omega})$-norm of both sides of the last equality, we get (2.6) for $j = 1$. We assume that (2.6) is already established for $j = l < r$ and prove (2.6) for $j = l + 1$. We write the equality

$$\begin{aligned} &F^{(l)}(\dot y + \dot\eta, y + \eta, x)(\dot h, h) - F^{(l)}(\dot y, y, x)(\dot h, h) - \big(F^{(l)}(\dot y, y, x)(\dot h, h)\big)^{(1)} \\ &\qquad = \int_0^1 (1-t)\big(F^{(l)}(\dot y + t\dot h, y + th, x)(\dot h, h)\big)^{(2)}(\dot\eta, \eta)\, dt. \end{aligned} \tag{2.7}$$

It is obvious that $A(*,*) = \big(F^{(l)}(\dot y, y, x)(\cdot,\cdot)\big)^{(1)}(*,*)$ belongs to the space

$$L\big(L((C^{k,\varepsilon}(\overline{\omega},\mathbb{R}^n))^l, C^{k-1,\varepsilon}(\overline{\omega})), C^{k-1,\varepsilon}(\overline{\omega})\big) \equiv X$$

with respect to the last two arguments. Since the $C^{k-1,\varepsilon}(\overline{\omega})$-norm of the right-hand side of (2.7) is estimated from above by $C|h|^l_{k,\varepsilon}|\eta|^2_{k,\varepsilon}$, the equality (2.7) implies

$$\big\|F^{(l)}[y+\eta] - F^{(l)}[y] - A[y]\eta\big\|_X \le C|\eta|^2_{k,\varepsilon}.$$

Thus, the derivative $F^{l+1}[y]$ exists and satisfies (2.6). Since differentiability implies continuity, the derivative $F^{(j)}[y]$, $j < r$, is continuous in y. To prove the continuity of the rth derivative, we use the equality

$$F^{(r)}(\dot y + \dot\eta, y + \eta, x)(\dot h, h) - F^{(r)}(\dot y, y, x)(\dot h, h)$$
$$= \int_0^1 (1-t)\big(F^{(r)}(\dot y + t\dot h, y + th, x)(\dot h, h)\big)^{(1)}(\dot\eta, \eta)\, dt.$$

Taking the $C^{k-1,\varepsilon}(\overline{\omega})$-norm of both sides of the last equality, we obtain

$$\Big\|F^{(r)}[y+\eta] - F^{(r)}[y]\Big\| \le C|\eta|_{k,\varepsilon},$$

where the norm of the left-hand side is taken in the space of r-linear mappings from $C^{k,\varepsilon}(\overline{\omega},\mathbb{R}^n)$ into $C^{k-1,\varepsilon}(\overline{\omega})$. □

We introduce the open set

$$V'_\rho = \{q \in \mathbb{R}^n : |q - z(x)| < \rho\}. \tag{2.8}$$

Let F be a function of $q \in V'_\rho$ and $x \in \overline{\omega}$ such that

$$\begin{gathered} F(q,x) \text{ has an } (r+1)\text{th order continuous derivative at } q \in \overline{V}'_\rho, \\ D_q^\beta F(q,x) \in C^{k,\varepsilon}(\overline{V}'_\rho \times \overline{\omega}), \quad |\beta| \le r+1. \end{gathered} \tag{2.9}$$

Under the above assumptions, the Taylor expansion

$$\begin{gathered} F(q+\delta q, x) = F(q,x) + \sum_{j=1}^{n} (j!)^{-1} F^{(j)}(q,x)(\delta q) \\ + (n!)^{-1} \int_0^1 F^{(n+1)}(q + t\delta q, x)(\delta q)(1-t)^n \, dt, \\ n = 1, \dots, r, \quad q, q+\delta q \in \overline{V}'_\rho, \end{gathered} \tag{2.10}$$

holds, where $F^{(j)}(q,x)(\delta q)$ denote terms containing $D_q^\beta F(q,x)$ for $|\beta| = j$. On the open sets

$$\begin{gathered} U'_\rho = \{y \in C^{k,\varepsilon}(\overline{\omega}, \mathbb{R}^n) : y(x) \in V'_\rho \text{ at } x \in \overline{\omega}\}, \\ U''_\rho = \{y \in C^{k,\varepsilon}(\partial\omega, \mathbb{R}^n) : y(x) \in V'_\rho \text{ at } x \in \partial\omega\} \end{gathered} \tag{2.11}$$

we define the mappings $F : U' \to C^{k,\varepsilon}(\overline{\omega})$ and $f : U'' \to C^{k,\varepsilon}(\partial\omega)$ by the equalities

$$F[y] = F(y(x), x), \quad f[y] = F(y(x), x). \tag{2.12}$$

LEMMA 2.2. *The mappings F and f have r continuous derivatives on the sets U'_ρ and U''_ρ respectively; moreover,*

$$F^{(j)}[y](h, \dots, h) = F^{(j)}(y(x), x)(h), \quad f^{(j)}[y](h, \dots, h) = F^{(j)}(y(x), x)(h).$$

The proof is similar to that of Lemma 2.1.

2.2. Differentiability of inverse operators. Let X be a Banach space. Consider an open set of invertible operators $U \subset L(X,X)$ and define the mapping $F : U \to L(X,X)$ by the equality

$$F[A] = A^{-1}, \quad A \in U. \tag{2.13}$$

LEMMA 2.3. *The mapping F is infinitely differentiable on U, and*

$$F^{(j)}[A](H, \dots, H) = (-1)^j (A^{-1}H)^j A^{-1}.$$

PROOF. Since X is a Banach space, for $A \in U$ and all $H \in L(X,X)$ small enough we have

$$\begin{aligned} (A-H)^{-1} - A^{-1} &= \sum_{j=1}^{\infty} (A^{-1}H)^j A^{-1} \\ &= \sum_{j=1}^{N} (A^{-1}H)^j A^{-1} + (A^{-1}H)^{N+1} (A-H)^{-1}. \end{aligned} \tag{2.14}$$

It is obvious that for every j

$$(2.15)\qquad \big(A^{-1}H\big)^j A^{-1} \in L\big(L(X,X)^j, L(X,X)\big).$$

The equality (2.14) for $N = 1$ takes the form

$$(A-H)^{-1} - A^{-1} - A^{-1}HA^{-1} = \big(A^{-1}H\big)^2 (A-H)^{-1}.$$

Since $\|(A^{-1}H)^2(A-H)^{-1}\| \le C\|H\|^2$ for H small enough, the mapping F is differentiable and $F'[A](H) = A^{-1}HA^{-1}$. Taking into account the form of $F'[A](H)$ and the differentiability of $F[A]$, we conclude that $F'[A](H)$ is differentiable; moreover, its derivative can be expressed as products of A^{-1} and H. Arguing as above, we can show that F is infinitely differentiable. From (2.14) we can derive the formula for the jth derivative. □

§3. Implicit function theorem

Let X, Y, and Z be Banach spaces, let $U \subset X \times Y$ be an open set containing the point $x = 0, y = 0$, and let $F : U \to Z$ be an r-times ($r \ge 1$) continuously differentiable mapping such that $F[0,0] = 0$. We consider

$$(3.1)\qquad F[x,y] = 0, \quad \|x\| \le \delta, \quad \|y\| \le \rho.$$

THEOREM 3.1 (on implicit function). *Let $F_y[0,0]$ be an isomorphism between the spaces Y and Z. There exist positive numbers ρ and δ such that any solution y to the problem* (3.1) *can be presented as follows*:

$$(3.2)\qquad y = u[x],$$

where u maps a ball of radius δ of the space X into a ball of radius ρ of the space Y, has r continuous derivatives there, and satisfies the conditions

$$(3.3)\qquad u[0] = 0, \quad u'[0] = -F_y^{-1}\big[x, u[x]\big] F_x\big[x, u[x]\big].$$

PROOF. We write the equation (3.1) in the following equivalent form:

$$Ay = Ay - F[x,y], \quad A = F_y[0,0], \quad \|x\| \le \delta, \quad \|y\| \le \rho.$$

Since the operator A is invertible, we have

$$(3.4)\qquad y = A^{-1}\big[Ay - F[x,y]\big] \equiv R[x,y], \quad \|x\| \le \delta, \quad \|y\| \le \rho.$$

We show that for sufficiently small ρ and δ the operator $R[x,y]$ maps a ball of radius ρ of the space Y into itself and that it is a contraction operator in this ball. Indeed,

$$\big\|R[x,y_1] - R[x,y_2]\big\|_Y \le \|A^{-1}\|_{L(Z,Y)} \big\|(F[x,y_1] - F[x,y_2]) - F_y[0,0](y_1 - y_2)\big\|_Z.$$

To estimate the last factor on the right-hand side of the last inequality, we apply the inequality (1.2) with $T = F_y[0,0]$ and obtain

$$\begin{aligned}\|F[x,y_1] &- F[x,y_2] - F_y[0,0](y_1-y_2)\big\|_Z \\ &\le \sup_{0\le t\le 1} \big\|F_y[x, ty_1 + (1-t)y_2] - F_y[0,0]\big\|_{L(Y,Z)} \big\|y_1 - y_2\big\|_Y.\end{aligned}$$

Taking ρ and δ small enough and using the continuity of the first derivative of F, we achieve the smallness of the first factor on the right-hand side of the estimate just obtained. Therefore, for sufficiently small ρ and δ we obtain

$$\big\|R[x,y_1]-R[x,y_2]\big\|_Y \le 2^{-1}\big\|y_1-y_2\big\|_Y, \quad \|x\|_Y \le \delta, \quad \|y_i\|_Y \le \rho, \quad i=1,2,$$

which yields

$$\|R[x,y]\|_Y \le \|R[x,y]-R[x,0]\|_Y + \|R[x,0]\|_Y \le 2^{-1}\|y\|_Y + \|R[x,0]\|_Y.$$

By the formula for R, we find that

$$\|R[x,0]\|_Y \le \|A^{-1}\|_{L(Z,Y)}\|F[x,0]\|_Z.$$

Taking into account the condition $F[0,0]=0$ and the continuity of F, and decreasing (if necessary) the number δ, we obtain $\|R[x,0]\|_Y \le 2^{-1}\rho$, Thus, for sufficiently small ρ and δ we have $\|R[x,y]\|_Y \le \rho$ for $\|x\|_X \le \delta$ and $\|y\|_Y \le \rho$.

By the contraction mapping principle, for each x such that $\|x\|_X \le \delta$ there exists a unique function y that solves the problem (3.4) and satisfies the estimate $\|y\|_Y \le \rho$. We write the solution y in the form (3.2). The point $x=0, y=0$ satisfies (3.4). Since the solution is uniquely determined, we conclude that $u[0]=0$.

We prove the continuity of u. Substituting the solution (3.2) in (3.4), we obtain the identity $u[x] \equiv R[x,u[x]]$, $\|x\| \le \delta$, from which for $\|x_i\| \le \delta$, $i=1,2$, we find that

$$\begin{aligned}\big\|u[x_1]-u[x_2]\big\|_Y &= \big\|R[x_1,u[x_1]]-R[x_2,u[x_2]]\big\|_Y \\ &\le \big\|R[x_1,u[x_1]]-R[x_1,u[x_2]]\big\|_Y + \big\|R[x_1,u[x_2]]-R[x_2,u[x_2]]\big\|_Y \\ &\le 2^{-1}\big\|u[x_1]-u[x_2]\big\|_Y + \big\|R[x_1,u[x_2]]-R[x_2,u[x_2]]\big\|_Y,\end{aligned}$$

which implies the estimate

$$\big\|u[x_1]-u[x_2]\big\|_Y \le 2\big\|R[x_1,u[x_2]]-R[x_2,u[x_2]]\big\|_Y, \quad \big\|x_i\big\|_X \le \delta.$$

Since $R[x,y]$ is continuous, the right-hand side of the last estimate tends to zero as $x_1 \to x_2$. Therefore, $u[x]$ is continuous.

Now we prove that $u[x]$ is differentiable. Substituting (3.2) in (3.1), we find that

$$F[x,u[x]] \equiv 0, \quad \|x\|_X \le \delta.$$

Let the norms of x_1 and x_2 be less than δ. We introduce the notation $x_1-x_2=\delta x$ and $u[x_1]-u[x_2]=\delta u$. Since F is differentiable, we have

$$F[x_2,u[x_2]]-F[x_1,u[x_1]]-F_y[x_1,u[x_1]]\delta u - F_x[x_1,u[x_1]]\delta x = \alpha(x_1,\delta x,\delta u),$$

where $\|\alpha\|_Z\left[\|\delta x\|_X+\|\delta u\|_Y\right]^{-1} \to 0$ as $\|\delta x\|_X+\|\delta u\|_Y \to 0$. The operator $F_y[0,0]$ is invertible. Note that $F_y[x,y]$ depends continuously on x and y, the mapping $u[x]$ is continuous, and $u[0]=0$. Hence we can conclude that for δ small enough the operator $F_y[x,u[x]]$ is invertible, $F_y^{-1}[x,u[x]]$ depends continuously on x, and the norm $\|F_y^{-1}[x,u[x]]\|_{L(Z,Y)}$ is uniformly bounded with respect to x. Therefore, we can write the above equality in the form

$$\delta u + B[x_1]\delta x = \beta(x_1,\delta x,\delta u), \tag{3.5}$$

where $B[x] = F_y^{-1}[x, u[x]]F_x[x, u[x]]$ and $\|\beta\|_Y \big[\|\delta x\|_X + \|\delta u\|_Y\big]^{-1} \to 0$ as $\|\delta x\|_X + \|\delta u\|_Y \to 0$. Taking into account the continuity of $u[x]$ and the obvious estimate

$$\big\|\delta u + B[x_1]\delta x\big\|_Y \le C\big[\|\delta x\|_X + \|\delta u\|_Y\big],$$

we conclude that $\|\beta\|_Y \big[\|\delta x\|_X + \|\delta u + B[x_1]\delta x\|_Y\big]^{-1} \to 0$ as $\|\delta x\|_X \to 0$. This, together with (3.5), implies

$$\|\delta u + B[x_1]\delta x\|_Y = o(\|\delta x\|_X) \text{ as } \delta x \to 0,$$

which means that $u[x]$ is differentiable at any point x_1 such that $\|x_1\|_X < \delta$ and the derivative satisfies (3.3). The continuity of the first order derivative and the existence of the 2nd order, ..., rth order continuous derivatives of $u[x]$ follow from the continuity and existence of the 1st order, ..., $(r-1)$th order derivatives of the right-hand side of (3.3). □

REMARK 3.1. Let numbers ρ and δ be such that for problem (3.1) Theorem 3.1 is valid. Then a solution to problem (3.1) is determined by (3.2). Let

$$\rho(\delta) = \sup \|u[x]\|_Y, \quad \|x\|_X \le \delta. \tag{3.6}$$

Then u takes the ball $B_\delta(0) \subset X$ to the ball $B_{\rho(\delta)}(0) \subset Y$. If $u[x]$ is defined in $B_\delta(0)$, it is also defined in any ball $B_{\delta'}(0)$, $\delta' < \delta$, and maps $B_{\delta'}(0) \subset X$ into $B_{\rho(\delta')}(0) \subset Y$. It is obvious that $\rho(\delta)$ monotonically decreases to zero as $\delta \to 0$.

REMARK 3.2. Let $\sup \|F_x[x, y]\| = C < \infty$, $x, y \in U$. Then δ can be chosen so that it will depend only on the constant C, the norm of the operator $F_y^{-1}[0, 0]$, and a majorant for the modulus of continuity

$$h(\varepsilon) = \sup \|F_y[x, y] - F_y[0, 0]\|, \quad \|x\| \le \varepsilon, \quad \|y\| \le \varepsilon, \quad \varepsilon > 0,$$

of the function $F_y[x, y]$ at the point $(0, 0)$.

THEOREM 3.2 (on inverse functions). *Let Y, X be Banach spaces, let $U \subset Y$ be a neighborhood of the origin, and let $F : U \to X$ be an r-times ($r \ge 1$) continuously differentiable mapping. Assume that $F[0] = 0$ and $F'[0]$ is an isomorphism between Y and X. Then there exist numbers δ and $\rho = \rho(\delta)$ such that the set of all solutions y to the problem $F[y] = x$, $\|y\| \le \rho$, $\|x\| \le \delta$, can be described by means of an r-times continuously differentiable function u so that $y = u[x]$ and $u[0] = 0$.*

PROOF. Theorem 3.2 immediately follows from Theorem 3.1 applied to the function $F[x, y] = F[y] - x$. Furthermore, as Z the space X appears here. □

REMARK 3.3. By Remark 3.2, we can choose δ so that it depends only on $\|F'[0]^{-1}\|$ and a majorant of the modulus of continuity $h(e)$ of the function $F'[y]$ at 0, where $h(\varepsilon) = \sup \|F'[y] - F'[0]\|$, $\|y\| \le \varepsilon$, $\varepsilon > 0$.

§4. Local structure of the set of solutions

Let X and Y be Banach spaces equipped with the norms $\|\cdot\|$ and $|\cdot|$ respectively, $U \subset X$ an open set containing 0, and $F : U \to Y$ an r-times continuously differentiable mapping. Assume that $F[0] = 0$ and the kernel N of the operator $F'[0]$ is not trivial, i.e., $N \ne \{0\}$. Consider the problem

$$F[x] = 0, \quad x \in U. \tag{4.1}$$

DEFINITION 4.1. We say that the set of all solutions to the problem (4.1) has the structure of a surface (or, for short, surface structure) at a point $x = 0$ if the space X can be decomposed into the direct sum

$$X = N \dot{+} N' \tag{4.2}$$

and there exists an r-times continuously differentiable mapping R from a neighborhood of zero and radius δ in the space N into a neighborhood of zero and radius $\rho = \rho(\delta)$ in the space N' such that any solution

$$x = u + v, \qquad u \in N, \quad \|u\| < \delta, \quad v \in N', \quad \|v\| < \rho, \tag{4.3}$$

to the problem (4.1) is represented in the form

$$x = u + R[u], \qquad u \in N, \quad \|u\| < \delta; \tag{4.4}$$

moreover, $R[0] = 0$ and $R'[0] = 0$.

THEOREM 4.1. *Suppose that for any $y \in Y$ the problem*

$$F'[0]w = y \tag{4.5}$$

has a solution that depends linearly on y and satisfies the estimate

$$\|w\| \leq C|y|. \tag{4.6}$$

Then the set of all solutions to the problem (4.1) *has surface structure at the point* $x = 0$.

PROOF. By assumption, there exists a bounded linear operator $T : Y \to X$ that takes any $y \in Y$ to a solution $w \in X$ to the problem (4.5). We define the operators P and Q by the equalities

$$Px = x - TF'[0]x, \quad Qx = TF'[0]x. \tag{4.7}$$

It is easy to see that P and Q are bounded projections on X, $P + Q = 1$, and $PX = N$. We decompose the space X into the direct sum

$$X = N \dot{+} N', \quad N' = QX, \tag{4.8}$$

and represent the solution x to the problem (4.1) in the form

$$x = u + v, \qquad u \in N, \quad v \in N', \quad u + v \in U.$$

Then the problem (4.1) takes the form

$$\varphi[v, u] \equiv F[v + u] = 0, \qquad u + v \in U \subset N \times N'. \tag{4.9}$$

Regarding φ as an r-times continuously differentiable mapping from U into Y, we note that

$$\varphi[0, 0] = 0, \quad \varphi_u[0, 0] = F'[0]|_N = 0, \quad \varphi_v[0, 0] = F'[0]|_{N'}.$$

By the above assumptions, $\varphi_v[0, 0]$ is a mapping from N' onto Y and is a one-to-one correspondence because $N \cap N' = \{0\}$. The continuity of the inverse mapping follows from the Banach theorem. Thus, $\varphi_v[0, 0]$ is an isomorphism between N' and Y. To complete the proof, it remains to apply the implicit function theorem to (4.9). □

REMARK 4.1. Substituting (4.4) in (4.1) and expanding the left-hand side of the identity $F[u + R[u]] \equiv 0$ in a Taylor series, we obtain recurrence relations for the differentials of $R[u]$ at zero. In particular, for $r \geq 2$

$$F'[0]R''[0](u,u) = -F''[0](u,u). \tag{4.10}$$

Together with the spaces X and Y, we consider one more pair of Banach spaces X_c and Y_c, equipped with the norms $\|\cdot\|_c$ and $|\cdot|_c$ respectively. Let $U_c \subset X_c$ be an open set containing the point 0 and let $F_c : U_c \to Y_c$ be an r-times continuously differentiable mapping. We suppose that

$$X \subset X_c, \quad \|\cdot\|_c \leq \alpha\|\cdot\|, \quad Y \subset Y_c, \quad |\cdot|_c \leq \alpha|\cdot|, \tag{4.11}$$

where the embeddings are assumed to be dense, and

$$U \subset U_c \cap X, \quad F_c\big|_U = F. \tag{4.12}$$

THEOREM 4.2. *If for any $y \in Y$ the problem* (4.5) *has a solution $w \in X$ that depends linearly on y and satisfies the estimates*

$$\|w\| \leq C|y|, \quad \|w\|_c \leq C|y|_c, \tag{4.13}$$

then the set of all solutions to the problem (4.1) *has surface structure at the point $x = 0$; moreover, $R[u]$, $\|u\| < \delta$, is an r-times continuously differentiable mapping from the normed space N equipped with the norm $\|\cdot\|_c$ to the normed space N' equipped with the same norm.*

PROOF. The fact that the set of all solutions to problem (4.1) has surface structure at $x = 0$ follows from Theorem 4.1. It remains to prove that R is an r-times continuously differentiable mapping from the space N into the space N' equipped with the norms equal to $\|\cdot\|_c$. By (4.13), the operator T that takes $y \in Y$ to a solution $w \in X$ of problem (4.5) admits an extension by continuity to the operator T_c that takes $y \in Y_c$ to a solution $w \in X_c$ of the problem $F_c'[0]w = y$. Hence $P_c = I_c - T_cF_c'[0]$ and $Q_c = T_cF_c'[0]$ are continuous extensions of the projections P and Q defined in (4.7); moreover, P_c projects X_c onto the kernel N_c of the operator $F_c'[0]$. We decompose the space X_c into the direct sum $X_c = N_c + N_c'$, where $N_c' = Q_cX_c$. Since the embedding of X into X_c is dense and

$$N = PX = P_cX \subset P_cX_c = N_c, \quad N' = QX = Q_cX \subset Q_cX_c = N_c',$$

it follows that the spaces N and N' are also embedded in N_c and N_c' respectively, and the embeddings are dense.

By Theorem 4.1, for sufficiently small δ_c and $\rho_c = \rho_c(\delta_c)$ any solution x to the problem

$$F_c[x] = 0, \quad x = \eta + \nu, \quad \eta \in N_c, \quad \nu \in N_c', \quad \|\eta\|_c \leq \delta_c, \quad \|\nu\|_c \leq \rho_c, \tag{4.14}$$

is represented in the form

$$x = \eta + R_c[\eta], \quad \eta \in N_c, \quad \|\eta\|_c \leq \delta_c, \tag{4.15}$$

where R_c is an r-times continuously differentiable mapping such that $R_c[0] = 0$ and $R_c'[0] = 0$. By the same theorem, for sufficiently small δ and $\rho = \rho(\delta)$ any solution x to the problem

$$F[x] = 0, \quad x = u + v, \quad u \in N, \quad v \in N', \quad \|u\| \leq \delta, \quad \|v\| \leq \rho, \tag{4.16}$$

is represented in the form

$$x = u + R[u], \qquad u \in N, \quad \|u\| \le \delta, \tag{4.17}$$

where R is an r-times continuously differentiable mapping such that $R[0] = 0$ and $R'[0] = 0$. We fix δ_c and choose δ so small as to satisfy the inclusion

$$W = \{x = u + v : u \in N, \|u\| \le \delta, v \in N', \|v\| \le \rho(\delta)\}$$
$$\subset \{x = \eta + \nu, \eta \in N_c, \|\eta\|_c \le \delta_c, \nu \in N'_c, \|\nu\|_c \le \delta_c(\rho_c)\}.$$

Then every solution $x \in W$ to (4.16) is simultaneously a solution to (4.14). Therefore, it admits both the representations (4.17) and (4.15), which implies that

$$u - \eta = R_c[\eta] - R[u]. \tag{4.18}$$

Since $N \subset N_c$ and $N' \subset N'_c$, the left-hand side of (4.18) belongs to N_c and the right-hand side of (4.18) belongs to N'_c. Therefore, $u = \eta$ and $R[u] = R_c[\eta]$. Thus,

$$R[u] = R_c[u], \qquad u \in N, \quad \|u\| \le \delta.$$

Then R_c, being a mapping from N into N' with norms equal to $\|\cdot\|_c$, has r continuous derivatives. The last equality shows that R possesses this property too. □

REMARK 4.2. Let $u, v \in N$ and $\|u\|$, $\|v\|$, $\|u+v\| \le \delta$. We expand the mapping $R[u+v]$ in a Taylor series at the point u and denote by $R^{\varkappa}[u]v$ the partial sum containing all the derivatives of order up to $\varkappa$. Then for $\varkappa \le r - 1$ we have

$$\|R[u+v] - R^{\varkappa}[u]v\| \le C\|v\|^{\varkappa+1}, \quad \|R[u+v] - R^{\varkappa}[u]v\|_c \le C\|v\|_c^{\varkappa+1}. \tag{4.19}$$

Indeed, the first estimate in (4.19) follows from the fact that the mapping $R[u]$, $u \in N$, $\|u\| \le \delta$, has r derivatives, and the second estimate is valid because $R_c[u]$ and $R[u]$ coincide for $u \in N$, $\|u\| \le \delta$, the mapping R_c has r derivatives, and the derivatives of R and R_c are equal (cf. 1.7).

We consider an analog of the problem (4.1) with an additional parameter. Let X, V, and Y be Banach spaces equipped with the norms $\|\cdot\|$, $\|\cdot\|'$, and $|\cdot|$ respectively, let $U \subset X \times Y$ be an open set containing zero, and let $F : U \to Y$ be an r-times continuously differentiable mapping. We assume that $F[0,0] = 0$ and the kernel N of the operator $F_x[0,0]$ is not trivial, i.e., $N \ne \{0\}$. We also assume that for any $y \in Y$ the equation $F_x[0,0]w = y$ has a solution that depends linearly on y and satisfies the estimate $\|w\| \le C|y|$. Then (cf. Theorem 4.1) the space X can be decomposed into the direct sum $X = N \dot{+} N'$.

Consider the problem

$$\begin{aligned} &F[x,v] = 0, \qquad x = \xi + \eta, \quad \xi \in N, \quad \eta \in N', \\ &\|\eta\| < \rho, \quad \|v\|' < \delta, \quad \|\xi\| < \delta. \end{aligned} \tag{4.20}$$

THEOREM 4.3. *There are numbers $\delta > 0$ and $\rho = \rho(\delta) > 0$ such that any solution to the problem* (4.20) *is represented in the form*

$$x = \xi + \eta[v, \xi], \tag{4.21}$$

where η is an r-times continuously differentiable mapping such that $\eta[0,0] = 0$ and $\eta_\xi[0,0] = 0$.

PROOF. Let $Z = N \times V$ be a Banach space equipped with the norm $\|\cdot\|''$ that equals the maximum over the norms of components. We write the problem (4.20) in the following equivalent form:

$$\Phi[\eta, z] = 0, \qquad \eta \in N', \quad \|\eta\| < \rho, \quad z \in Z, \quad \|z\|'' < \delta, \tag{4.22}$$

and apply the implicit function theorem to the problem (4.22). □

REMARK 4.3. If we want to define an r-times continuously differentiable solution $x = x[v]$, $\|v\|' < \delta$, to the equation $F[x, v] = 0$ that satisfies the condition $x[0] = 0$, then it suffices to substitute an arbitrary r-times continuously differentiable mapping $\xi = \xi[v]$ such that $\|\xi[v]\| < \delta$ and $\xi[0] = 0$ in (4.21).

§5. The Lyapunov–Schmidt splitting procedure

In this section, we return to the problem (4.1), but under some other assumptions on solvability of the linearized problem (4.5).

Let spaces X, Y, X_c, Y_c and mappings F, F_c be the same as in §4 and let l be a bounded linear functional on Y_c. By (4.11), l is also bounded on Y. We fix $e \in Y$ such that $l(e) = 1$ and write the problem (4.1) in the following (equivalent) form:

$$\Phi[x] \equiv F[x] - el(F[x]) = 0, \quad l(F[x]) = 0, x \in U. \tag{5.1}$$

We give a local description of the set of all solutions to the first equation in (5.1).

THEOREM 5.1. *Suppose that the problem* (4.5) *is solvable for those and only those $y \in Y$ for which*

$$l(y) = 0. \tag{5.2}$$

Let (5.2) *hold and let the problem* (4.5) *have a solution $w \in X$ that depends linearly on y and satisfies the estimates* (4.6) *and* (4.13). *Then the set of all solutions to the equation*

$$\Phi[x] = 0, \quad x \in U, \tag{5.3}$$

has surface structure at the point $x = 0$; moreover, $R[u]$, $\|u\| < \delta$, is an r-times continuously differentiable mapping from the space N equipped with the norm $\|\cdot\|_c$ into the space N' equipped with the same norm $\|\cdot\|_c$.

PROOF. The operator $Sy = el(y)$ is a bounded projection in Y and Y_c. Its range is one-dimensional and is spanned on the vector e. Since the operator $(1 - S)y = y - el(y)$, is the complementary projection, it is bounded in Y and Y_c. The images of Y and Y_c under this projection will be denoted by Y' and Y'_c respectively. It is obvious that Y' is dense in Y'_c, the mappings $\Phi : U \to Y'$ and $\Phi_c = F_c - l(F_c[\cdot]) : U_c \to Y'_c$ are r-times continuously differentiable, and

$$\Phi'[0]w = F'[0]w - el(F'[0]w) = F'[0]w.$$

Under the above assumptions, for every $y \in Y'$ the equation $\Phi'[0]w = y$ has a solution $w \in X$ that depends linearly on y and satisfies the estimates (4.6) and (4.13). Therefore, we can apply Theorem 4.2 to the problem (5.3). □

Thus, any solution to the first equation in (5.1) is represented in the form (4.4). Therefore, a function x of the form (4.4) is a solution to the problem (4.1) for those and only those u that satisfy the bifurcation equation

$$\Psi[u] \equiv l(F[u + R[u]]), \qquad u \in N, \quad \|u\| < \delta. \tag{5.4}$$

The following assertion will be useful in what follows.

LEMMA 5.1. *The function $\Psi[u]$, $u \in N$, $\|u\| < \delta$, regarded as a function on the space N equipped with the norm $\|\cdot\|$ and as a function on N equipped with the norm $\|\cdot\|_c$, has r continuous derivatives. The derivatives of $\Psi[u]$ are independent of the choice of the norm in N. In addition, the following conditions are fulfilled:*

$$\Psi[0] = 0, \quad \Psi'[0] = 0, \quad \Psi''[0](u, u) = l(F''[0](u, u)). \tag{5.5}$$

PROOF. The existence of r continuous derivatives of $\Psi[u]$, $u \in N$, $\|u\| < \delta$, follows from the fact that $F[x]$, $F_c[x]$, and $R[u]$ are r-times continuously differentiable. The independence of the derivatives of $\Psi[u]$ from the choice of the norm in N is a consequence of Remark 4.2 and Subsection 1.7. Since $F[0] = 0$ and $R[0] = 0$, we have $\Psi[0] = 0$. Furthermore,

$$\Psi[u] - \Psi[0] = l(F'[0]u) + \alpha(u), \quad |\alpha(u)|\,\|u\|^{-1} \to 0 \text{ as } \|u\| \to 0.$$

By the assumption (5.2), we have $l(F'[0]u) = 0$. Hence $\Psi'[0] = 0$. Similarly,

$$\Psi[u] - \Psi[0] = l(F'[0]u) + 2^{-1}(F''[0](u, u)) + 2^{-1}l(F'[0]R''[0](u, u)) + \alpha(u),$$
$$|\alpha(u)|\,\|u\|^{-2} \to 0 \text{ as } \|u\| \to 0.$$

Using (5.2) once more, we obtain the required formula for $\Psi''[0](u, u)$. □

REMARK 5.1. Let u, $v \in N$ and $\|u\|$, $\|v\|$, $\|u+v\| < \delta$. Denote by $\Psi^{(\varkappa)}[u]v$ the finite sum of the Taylor series of the mapping $\Psi[u+v]$ at the point u that contains derivatives of order up to $\varkappa$. Since $\Psi[u]$ is r-times continuously differentiable and the derivatives are independent of the choice of the norm in N, for all $\varkappa \le r - 1$ we obtain the estimates

$$\left|\Psi[u + v] - \Psi^{(\varkappa)}[u]v\right| \le C_1\|v\|_c^{\varkappa+1} \le C_2\|v\|^{\varkappa+1}. \tag{5.6}$$

We proceed to the proof of the main result of this section.

THEOREM 5.2. *Let the assumptions of Theorem 5.1 hold, and let $r \ge 3$.*

1. *If the quadratic form $\Psi''[0](u, u)$ is indefinite on N, then $x = 0$ is a nonisolated solution of class X to the problem (4.1).*

2. *Assume that it is possible to introduce a norm $\langle\cdot\rangle$ in N so that the inequality*

$$|\Psi''[0](u, u)| \ge C\langle u\rangle^2, \quad C > 0, \tag{5.7}$$

and the multiplicative inequality

$$\|u\|_c^3 \le \alpha\langle u\rangle^2\|u\|, \quad u \in N, \quad \alpha > 0, \tag{5.8}$$

hold. Then $x = 0$ is a nonisolated solution to the problem (4.1) in X.

PROOF. (a) Since the quadratic form $\Psi''[0](u,u)$ is indefinite, there are $u_1 \in N$ and $u_2 \in N$ such that

$$\Psi''[0](u_1,u_1) > 0, \quad \Psi''[0](u_2,u_2) < 0.$$

We choose a constant $\mu > 0$ so as to satisfy the inequalities

$$\Psi''[0](u_1,u_1) > \mu\|u_1\|^2, \quad \Psi''[0](u_2,u_2) < 0.$$

These inequalities remain valid after replacing u_i by $v_i = \lambda u_i$, $\lambda \neq 0$. From (5.6) with $\varkappa = 0$, $u = 0$, $v = v_i$, $i = 1, 2$, and small $|\lambda| \neq 0$, we find that

$$\Psi[v_1] \geq 2^{-1}\Psi''[0](v_1,v_1) - C_1\|v_1\|^3 > \|v_1\|^2(\mu/2 - C_1\|v_1\|) > 0,$$
$$\Psi[v_2] \leq 2^{-1}\Psi''[0](v_2,v_2) + C_1\|v_2\|^3 < -\|v_2\|^2(\mu/2 - C_1\|v_2\|) < 0.$$

Since v_1 and v_2 are linearly independent for $\lambda \neq 0$, the interval $v(t) = tv_1 + (1-t)v_2$, $t \in [0,1]$, does not contain the origin of N. By the continuity of Ψ, there exists $t_* = t_*(\lambda) \in (0,1)$ such that $\Psi[v(t_*)] = 0$. Since $t_*(\lambda) \to 0$ for $\lambda \to 0$, for small $|\lambda| \neq 0$ the family $v(t_*(\lambda))$ is constituted by nonzero solutions to the problem $\Psi[v] = 0$; moreover, $\|v(t_*(\lambda))\| \to 0$ as $\lambda \to 0$. Therefore, for small $|\lambda| \neq 0$ the family $x(\lambda) = v(t_*(\lambda)) + R[v(t_*(\lambda))]$ consists of nonzero solutions to the problem (4.1), and $\|x(\lambda)\| \to 0$ as $\lambda \to 0$. The existence of such a family $x(\lambda)$ means that the solution $x = 0$ to the problem (4.1) is nonisolated in X.

(b) Let x be a solution to the first equation in (5.1). The solution x admits the representation (4.4). To prove the required assertion, it suffices to establish the inequality

$$|\Psi[u]| \geq C\|x\|_c^3, \quad C > 0. \tag{5.9}$$

Taking (5.6) with $u = 0$, $\varkappa = 2$, and u instead of v, we obtain

$$|\Psi[u]| \geq 2^{-1}|\Psi''[0](u,u)| - C'\|u\|_c^3, \qquad u \in N, \quad \|u\| < \delta, \quad C' > 0.$$

From the last inequality, the assumption (5.7), and the multiplicative inequality (5.8) it follows that

$$|\Psi[u]| \geq 2^{-1}C\langle u\rangle^2 - C'\|u\|_c^3 \geq \langle u\rangle^2[2^{-1}C - \alpha C'\|u\|], \qquad u \in N, \quad \|u\| < \delta.$$

Therefore, $|\Psi[u]| \geq \alpha\langle u\rangle^2$ for δ small enough. Multiplying it by $\|u\|$ and using the multiplicative inequality (5.8), we arrive at the inequality

$$\delta|\Psi[u]| \geq \|u\|\,|\Psi[u]| \geq \alpha^{-1}\|u\|_c^3. \tag{5.10}$$

From the second inequality in (4.19) with $u = 0$, $\varkappa = 1$, and $v = u$ and the equality $R'[0] = 0$ we obtain $\|R[u]\|_c \leq C\|u\|_c^2$. Hence

$$\|u\|_c \geq \|u + R[u]\|_c - \|R[u]\|_c \geq \|x\|_c - C\|u\|_c^2 \geq \|x\|_c - C'\|u\|\,\|u\|_c$$

for $u \in N$, $\|u\| < \delta$, which yields the inequality $\|u\|_c \geq C\|x\|_c$. Combining it with (5.10), we obtain (5.9). □

§6. The Morse lemma and the bifurcation problem

In this section, we study some properties of solutions to the problem (4.1) with F depending on a real parameter λ. We begin with some preliminary constructions. Let a function $f(v)$ be p-times ($p \geq 3$) continuously differentiable in a neighborhood U of the origin of the space $\mathbb{R}^n$, $n \geq 2$. Assume that

$$\begin{gathered} f(0) = 0, \quad f_v(0) = 0, \\ f_{vv}(0) \text{ is a nonsingular } n \times n \text{ matrix.} \end{gathered} \tag{6.1}$$

LEMMA 6.1 (Morse Lemma). *If $\delta > 0$ is small enough, then*

$$f(v) = 2^{-1} f_{vv}(0)(y(v), y(v)), \tag{6.2}$$

where $y(v)$, $|v| \leq \delta$, is some $(p-2)$-times continuously differentiable mapping that takes its values in some neighborhood of the origin of $\mathbb{R}^n$ and satisfies the conditions

$$y(0) = 0, \quad y_v(0) = 1. \tag{6.3}$$

PROOF. Using the Newton–Leibniz formula and integrating by parts, we obtain

$$f(v) = \left[\int_0^1 f_{vv}(tv)(1-t)\, dt \right](v, v). \tag{6.4}$$

The doubly symmetric $n \times n$ matrix in brackets on the right-hand side of (6.4), denoted by $B(v)$ below, has $p-2$ continuous derivatives in the domain and satisfies the condition $B(0) = f_{vv}(0)$. In terms of the matrix B, the problem (6.3) can be written as follows:

$$(B(v)v, v)_n = \big(f_{vv}(0)y(v), y(v)\big)_n. \tag{6.5}$$

We seek y in the form

$$y(v) = A(v)v, \tag{6.6}$$

where the $(p-2)$-times continuously differentiable $(n \times n)$-matrix-valued function $A(v)$ is defined in a neighborhood of the origin of $\mathbb{R}^n$ and satisfies the condition $A(0) = 1$. Since $p \geq 3$, the last property of the matrix A and (6.6) show that (6.2) is valid for $y(v)$ defined in (6.6). The equality (6.5) holds provided that $A(v)$ is such that

$$A^*(v) f_{vv}(0) A(v) - B(v) = 0. \tag{6.7}$$

We define the mapping Ψ by the equality

$$\Psi[A, v] = A^* f_{vv}(0) A - B(v).$$

It is obvious that Ψ is defined in some neighborhood of the point $(1, 0)$ of the space that is the direct product of the space of $n \times n$ matrices and the space $\mathbb{R}^n$. In addition, Ψ has $p-2$ continuous derivatives at $(1, 0)$ and acts in the space of symmetric $n \times n$ matrices. The problem of finding a matrix-valued function $A(v)$ satisfying (6.7) can be thought as the problem of solving the equation

$$\Psi[A, v] = 0, \qquad |A - 1| < s, \quad |u| < \delta, \tag{6.8}$$

with respect to A. We note that

$$\Psi[1,0]=0, \quad \Psi_A[1,0]h = h^* f_{vv}(0) + f_{vv}(0)h.$$

Moreover, the equation $\Psi_A[1,0]h = H$ is solvable for every symmetric matrix H. For a solution we can take the matrix $h = 2^{-1} f_{vv}^{-1}(0)H$. Therefor, we can apply Theorem 4.3 and Remark 4.3 to the problem (6.8). □

We use Lemma 6.1 to describe the set of all solutions to the problem

$$f(v)=0, \quad |v| \le \rho, \tag{6.9}$$

for ρ small enough and a p-times ($p \ge 3$) continuously differentiable function f satisfying (6.1). Using (6.3), we write (6.9) in the form

$$f_{vv}(0)(y(v), y(v)) = 0. \tag{6.10}$$

From the equation (6.1), which is quadratic with respect to y, we find y. After that we find v from y. The last operation is legitimate due to the condition (6.2) and the inverse function theorem.

We discuss a special case which will be important in the sequel. Let $n=2$ and let the determinant of the matrix $f_{vv}(0)$ be negative. Choosing a basis for $\mathbb{R}^2$ in which the matrix $f_{vv}(0)$ is diagonal, we rewrite (6.10) as follows:

$$\lambda_1 y_1^2 - \lambda_2 y_2^2 = 0, \tag{6.11}$$

where $\lambda_i > 0$, $i = 1,2$, and y_i are the coordinates of y in the above basis. The set of all solutions to the problem (6.11) can be graphically depicted by two lines intersecting at some angle. Therefore, for ρ small enough the set of all solutions to the problem (6.9) can be depicted by two curves intersecting at the origin, and the angle between these curves is the same as that between the above mentioned lines.

Now we turn to the bifurcation problem. Let X and Y be Banach spaces, $U \subset X \times \mathbb{R}$ a neighborhood of the origin, $F : U \to Y$ an r-times continuously differentiable mapping, and

$$F[0,\lambda] = 0, \quad \lambda \in U \cap \mathbb{R}. \tag{6.12}$$

DEFINITION 6.1. A point $\lambda = 0$ is called a bifurcation point for the problem

$$F[x,\lambda] = 0, \quad (x;\lambda) \in U, \tag{6.13}$$

if, in any neighborhood of the origin of the space $X \times \mathbb{R}$, there exists a solution $(x;\lambda)$ to the problem (6.18) such that $x \ne 0$.

LEMMA 6.2. *If $\lambda = 0$ is a bifurcation point, then the linear operator $F_x[0,0] : X \to Y$ is not an isomorphism between X and Y.*

PROOF. On the contrary, assume that the operator $F_x[0,0]$ is an isomorphism between X and Y. By the implicit function theorem, the problem (6.13) has a unique local solution $x = x(\lambda)$ and, by (6.12), $x(\lambda) \equiv 0$, i.e., $\lambda = 0$ is not a bifurcation point. □

The necessary condition for bifurcation obtained in Lemma 6.2 is not a sufficient one. The following example confirms this assertion: the system of equations

$$\lambda x_1 = x_2^3, \quad \lambda x_2 = -x_1^3$$

has only the zero solution for all λ and corresponds to the problem (6.13) with $X = Y = \mathbb{R}^2$. We give sufficient conditions for bifurcation.

THEOREM 6.1. *Let a function F have $r \geq 3$ continuous derivatives in U and satisfy the following conditions:*

(a) *the assumptions of Theorem 5.1 hold for the problem $F_x[0,0]w = y$,*

(b) *there exists a vector u_0 of the kernel of the operator $F_x[0,0]$ such that $l(F_{x\lambda}[0,0]u_0) \neq 0$.*

Then $\lambda = 0$ is a bifurcation point.

PROOF. Let $\widehat{X} = X \times \mathbb{R}$, $\widehat{x} \in \widehat{X}$, and $\widehat{x} = (x;\lambda)$, $x \in X$, $\lambda \in \mathbb{R}$. We define the r-times continuously differentiable mapping $\widehat{F} : U \to Y$ by the equality $\widehat{F}[\widehat{x}] = F[x,\lambda]$. Then

$$\widehat{F}[\widehat{0}] = 0, \quad \widehat{F}'[\widehat{0}]\widehat{h} = F_x[0,0]h + F_\lambda[0,0]\mu, \quad \widehat{h} = (h;\mu),$$
$$\widehat{F}''[\widehat{0}](\widehat{h},\widehat{h}) = F_{xx}[0,0](h,h) + 2F_{x\lambda}[0,0](h,\mu) + F_{\lambda\lambda}[0,0](\mu,\mu).$$

By (6.12), we have $F_\lambda[0,0] = 0$ and $F_{\lambda\lambda}[0,0] = 0$. Consequently, the kernel $\widehat{N}$ of the operator $\widehat{F}'[\widehat{0}]$ is given by $\widehat{N} = N \times \mathbb{R}$, and the assumptions of Theorem 5.1 hold for the problem $\widehat{F}'[\widehat{0}]\widehat{w} = y$ (for the required solution $\widehat{w}$ we can take $\widehat{w} = (w;0)$, where w is a solution to the problem $F_x[0,0]w = y$ whose existence is guaranteed by condition (a) of the theorem).

We write the problem (6.13) in the form $\widehat{F}[\widehat{x}] = 0$. Using the equivalent problem similar to (5.1), we find

$$\widehat{\Phi}[\widehat{x}] \equiv \widehat{F}[\widehat{x}] - el(\widehat{F}[\widehat{x}]) = 0, \quad l(\widehat{F}[\widehat{x}]) = 0, \quad \widehat{x} \in U. \tag{6.14}$$

By Theorem 5.1, the set of all solutions to the first equation in (6.14) has surface structure:

$$\widehat{x} = \widehat{u} + R[\widehat{u}], \quad \widehat{u} \in \widehat{N}, \quad \|\widehat{u}\| \leq \delta. \tag{6.15}$$

Compared to Definition 4.1, the mapping $\widehat{R}$ has the following additional property: $\widehat{R}[\widehat{u}] = 0$ for $\widehat{u} = (0;\lambda)$. Substituting this representation in the second equation in (6.14), we obtain the equation for $\widehat{u}$ such that (6.15) is a solution to (6.14), i.e.,

$$\widehat{\Psi}[\widehat{u}] \equiv l\big(\widehat{F}[\widehat{u} + \widehat{R}[\widehat{u}]]\big) = 0, \quad \widehat{u} \in \widehat{N}, \quad \|\widehat{u}\| \leq \delta. \tag{6.16}$$

By Lemma 5.1,

$$\widehat{\Psi}[\widehat{0}] = 0, \quad \widehat{\Psi}'[\widehat{0}] = 0, \quad \widehat{\Psi}''[\widehat{0}](\widehat{u},\widehat{u}) = l\big(F''[\widehat{0}](\widehat{u},\widehat{u})\big). \tag{6.17}$$

Let $u_0 \in N$ satisfy (b). Denote by N_0 the subspace of N spanned by u_0. Since N_0 is a one-dimensional space, there is a complement to N_0 in N. Hence $N = N_0 \dot{+} N_0'$. For $N_0 = N$ we set $N_0' = \{0\}$. Taking into account the above decomposition of $\widehat{N}$, we have $\widehat{N} = V \times W$, $V = N_0 \times \mathbb{R}$, $W = N_0'$. We rewrite (6.16) as follows:

$$\begin{gathered} f(v,w) \equiv \widehat{\Psi}[\widehat{u}] = 0, \quad \widehat{u} = (u;\lambda), \quad u \equiv cu_0 + u', \\ u' \in N_0', \quad v = (c;\lambda), \quad w = u'; \end{gathered} \tag{6.18}$$

the function $f(v, w)$ has r continuous derivatives in a neighborhood of the origin of $V \times W$ and, in view of (6.17), satisfies the relations

$$f(0,0) = 0, \quad f_v(0,0) = 0, \quad f_{vv}(0,0) = \begin{bmatrix} l(F_{xx}[0,0](u_0,u)), & l(F_{x\lambda}[0,0]u_0) \\ l(F_{x\lambda}[0,0]u_0), & l(F_{\lambda\lambda}[0,0]) \end{bmatrix}.$$

Since $F_{\lambda\lambda}[0,0] = 0$ and $l(F_{x\lambda}[0,0]u_0) \neq 0$, we have $\det f_{vv}(0,0) < 0$. Hence the set of all solutions to the problem (6.18) for $w = 0$ can be graphically presented as two curves lying in V and intersecting at a nonzero angle at the origin. The line $v = (0; \lambda)$ is one of these curves. Therefore, on the second curve, there exist points that are arbitrarily close to the origin and have a nonzero component c. Thus, $\lambda = 0$ is a bifurcation point for the problem (6.13). □

§7. Local structure of the set of solutions

Let $\omega \subset \mathbb{R}^m$ be a bounded domain with boundary $\partial\omega \in C^{k+1,\varepsilon}$, where k is a natural number and $\varepsilon \in (0,1]$, let $y(x)$ be a mapping of ω into $\mathbb{R}^n$, and let $\dot{y}(x)$ be the Jacobi matrix of $y(x)$. We fix a function $z \in C^{k,\varepsilon}(\overline{\omega}, \mathbb{R}^n)$ and, on the sets V_ρ and V'_ρ (cf. (2.1) and (2.8)), define functions $F(p,q,x)$, $p, q \in V_\rho$, $x \in \overline{\omega}$, and $\varphi(q,x)$, $q \in V'_\rho$, $x \in \overline{\omega}$, satisfying (2.2) and (2.9) respectively for some natural number r.

Denote by $\Gamma \subset \partial\omega$ an open set (the cases $\Gamma = \varnothing$ and $\Gamma = \partial\omega$ are possible) and put $S = \partial\omega/\overline{\Gamma}$. We assume that there is a finite number of connected components of $\partial\Gamma$ and every connected component of $\partial\Gamma$ is of class $C^{k,\varepsilon}$.

In ω, we consider the equation

$$F[y] \equiv F(\dot{y}(x), y(x), x) = 0, \quad y \in U_\rho, \tag{7.1}$$

where the set U_ρ is defined in (2.4). We fix a collection $\overline{x}$ of $x_j \in \omega$, $j = 1, \dots, l$, and impose the following conditions:

$$y(x) - z(x) = 0, \quad x \in \Gamma \cup \overline{x}, \qquad \varphi[y] \equiv \varphi(y(x), x) = 0, \quad x \in S. \tag{7.2}$$

The cases in which the conditions (7.2) are absent on Γ, S, or at points of $\overline{x}$ are possible. Such "shorter" conditions will be referred as to the conditions (7.2) but with $*$ instead of Γ, S, or $\overline{x}$ respectively.

We assume that $z(x)$ satisfies (7.1) and (7.2), i.e.,

$$F[z] = 0, \quad x \in \omega, \qquad \varphi[z] = 0, \quad x \in \partial\omega. \tag{7.3}$$

We want to describe the set of all solutions to the problem (7.1), (7.2) for ρ small enough under some additional assumptions on the functions F, φ, and z. Introduce the notation

$$\begin{gathered} A_{ij}(x) = F_{p^i_j}(\dot{z}(x), z(x), x), \quad x \in \omega, \\ a^i(x) = -\frac{d}{dx_j} F_{p^i_j}(\dot{z}(x), z(x), x) + F_{q^i}(\dot{z}(x), z(x), x), \quad x \in \omega, \\ \gamma^i(x) = \varphi_{q^i}(z(x), x), \quad x \in \partial\omega, \\ Lu = \operatorname{div} A^* u + (a, u)_n, \quad u \in C^{k,\varepsilon}(\overline{\omega}, \mathbb{R}^n). \end{gathered} \tag{7.4}$$

Assume that

$$\begin{gathered} A_{ij} \in C^{k,\varepsilon}(\overline{\omega}), \quad a^i \in C^{k,\varepsilon}(\overline{\omega}), \quad \gamma^i \in C^{k,\varepsilon}(\partial\omega), \\ \det G(x) \neq 0, \quad G(x) = A^*(x)A(x), \quad |\gamma(x)| \equiv 1, \quad x \in \partial\omega. \end{gathered} \tag{7.5}$$

In particular, (7.5) implies that the operator L in (7.4) satisfies (1.1), (1.2) from Chapter 1. The operator L^* formally adjoint to L is given by the equality $L^*q = -A\nabla q + aq$.

We pass to constructions that allow us to write the problem (7.1), (7.2) as an abstract equation studied in §§4 and 5. We introduce the spaces

$X = \{h \in C^{k\varepsilon}(\overline{\omega}, \mathbb{R}^n) : h(x) = 0, x \in \Gamma u \overline{x}\}$,

$Z = \varnothing$ if $\Gamma = \partial\omega$ or $S = *$,

$Z = C^{k,\varepsilon}(\overline{S})$ if $* = \Gamma \neq \partial\omega, S \neq *$,

$Z =$ restrictions to S of functions of class $C^{k,\varepsilon}$ vanishing on Γ if $\Gamma \neq \varnothing, \partial\omega, *$, $S \neq *$,

$Y = C^{k-1,\varepsilon}(\overline{\omega}) \times Z$.

Let $h(x) = y(x) - z(x)$ and $U = \{h \in X : |h|_{k,\varepsilon} < \delta\}$. We define the mapping $T : U \to Y$ by the equality

$$T[h] = \{F[y], \varphi[y]\big|_S\}. \tag{7.6}$$

Under the above assumptions on F and φ, for δ small enough the mapping T is well defined in U and, by Lemmas 2.1 and 2.2, has r continuous derivatives there; moreover,

$$\begin{aligned} T'[0]h &= \{Lh, (h,\gamma)_n\big|_S\}, \\ T''[0](h,h) &= \{F''[z](h,h), \varphi''[z](h,h)\big|_S\} \end{aligned} \tag{7.7}$$

for $r \geq 1$ and $r \geq 2$ respectively, for all $h \in X$. The space $N = J^{k,\varepsilon}(\Gamma, S, \overline{x})$ is the kernel of the operator $T'[0]$:

$$J^{k,\varepsilon}(\Gamma, S, \overline{x}) = \{u \in C^{k,\varepsilon}(\overline{\omega}, \mathbb{R}^n) : u\big|_{\Gamma\cup\overline{x}} = 0, (u,\gamma)_n\big|_S = 0, Lu = 0, x \in \omega\}. \tag{7.8}$$

The problem (7.1), (7.2) of seeking a solution $y(x)$ lying in a sufficiently small $C^{k,\varepsilon}(\overline{\omega}, \mathbb{R}^n)$-neighborhood of the solution $z(x)$ can be rewritten in the terms of the mapping T as follows:

$$T[h] = 0, \quad h \in U. \tag{7.9}$$

THEOREM 7.1. 1. *Suppose that the equation* $L^*q = 0$, $q \in C^1(\overline{\omega})$, *either has only the zero solution or has a nonzero solution but one of the following conditions holds*:

(a) $\Gamma \neq \partial\omega$, $S = *$,

(b) $\Gamma = *$, $S \neq \partial\omega$,

(c) $* \neq \Gamma \neq \partial\omega$, $S \neq *$, $A(x_0)\nu(x_0) \neq \beta\gamma(x_0)$ *for some* $x_0 \in S$ *and all* $\beta \in \mathbb{R}$.

Then the set of all solutions to the problem (7.9) *has surface structure at the point* $h = 0$, *i.e., if a solution* $y(x)$ *to the problem* (7.1), (7.2) *is sufficiently close to* $z(x)$ *in the* $C^{k,\varepsilon}(\overline{\omega}, \mathbb{R}^n)$-*norm, then it is represented in the form*

$$y = z + u + R[u], \quad u \in N, \quad |u|_{k,\varepsilon} < \delta, \quad R[0] = 0, \quad R'[0] = 0, \tag{7.10}$$

where $R[u]$ *is an* r-*times continuously differentiable mapping.*

2. *Denote by* $R^{\varkappa}[u]v$ *the finite sum of the Taylor series of the mapping* $R[u+v]$ *at* u *that contains the derivatives of order up to* $\varkappa$. *Then for* $\varkappa \leq r-1$

$$\big|R[u+v] - R^{\varkappa}[u]v\big|_{s,\varepsilon} \leq C\|v\|_{s,\varepsilon}^{\varkappa+1}, \quad s = 1, \dots, k. \tag{7.11}$$

3. *For all $u \in N$ the second order derivative of R at the origin satisfies the equality*

$$\begin{aligned} LR''[0](u,u) &= -F''[z](u,u), \quad x \in \omega, \\ R''[0](u,u) &= 0, \quad x \in \Gamma \cup \overline{x}, \\ (R''[0](u,u),\gamma)_n &= -\varphi''[z](u,u), \quad x \in S. \end{aligned} \tag{7.12}$$

PROOF. (a) By Theorem 4.1, it suffices to show that for any $y \in Y$ the equation $T'[0]w = y$ has a solution $w \in X$ that depends linearly on the right-hand side and satisfies the estimate (4.6), i.e., for any pair $y = \{f;\chi\} \in Y$ the problem

$$Lw = f, \quad w(x) = 0, \quad x \in \Gamma \cup \overline{x}, \quad (w,\gamma)_n = \chi, \quad x \in S, \tag{7.13}$$

has a solution $w \in C^{k,\varepsilon}(\overline{\omega}, \mathbb{R}^n)$ that depends linearly on the pair $\{f,\chi\}$ and satisfies the estimate

$$|w|_{k,\varepsilon} \le C\big[|f|_{k-1,e} + |\chi|_{k,\varepsilon}(S)\big] \tag{7.14}$$

(for $S = *$ the last summand on the right-hand side of (7.14) vanishes). We extend the function $\chi(x)$ to the entire boundary $\partial\omega$ by zero in the cases $\Gamma \neq \varnothing, \partial\omega, *$ and $S \neq *$, or with the help of the extension operator Π (cf. the Appendix) in the cases $* = \Gamma \neq \partial\omega$ and $S \neq *$ (see the definition of the space Z). The extended function will be denoted by $\widehat{\chi}(x)$. A solution to the problem (7.13) can be found from the conditions

$$Lw = f, \quad w|_\Gamma = \varphi|_\Gamma, \quad (w,\gamma)_n\big|_S = (\varphi,\gamma)_n\big|_S, \quad w(x_j) = 0, \quad x_j \in \overline{x}, \tag{7.15}$$

with the vector-valued function $\varphi(x) = \gamma(x)\widehat{\chi}(x)$. The solvability of (7.15), linear dependence of the solution on the pair $\{f,\chi\}$, and the estimate (7.14) follow from Theorem 5.2 (Chapter 1).

(b) This assertion is a consequence of Theorem 4.2 and Remark 4.2. For X_c we take the closure of X in the $C^{s,\varepsilon}(\overline{\omega}, \mathbb{R}^n)$-norm and for Y_c we take the closure of Y in the $C^{s-1,\varepsilon}(\overline{\omega}) \times C^{s,\varepsilon}(\overline{S})$-norm, where $1 \le s \le k$. The estimate (4.13) for a solution to the problem (4.5) coincides with the estimate

$$|w|_{s,\varepsilon} \le C\big[|f|_{s-1,\varepsilon} + |\chi|_{s,\varepsilon}(s)\big] \tag{7.16}$$

for a solution to the problem (7.13). As in (a), we reduce (7.13) to (7.15) and apply Theorem 5.2 from Chapter 1.

(c) This assertion immediately follows from formulas (4.10), (7.13) and Lemmas 2.1 and 2.2. □

We study the case in which the problem $L^*q = 0$, $q \in C^1(\overline{\omega})$, has a nonzero solution but none of the conditions (a)–(c) of Theorem 7.1 holds.

LEMMA 7.1. *Suppose that the problem $L^*q = 0$, $q \in C^1(\overline{\omega})$, has a nonzero solution and one of the following conditions holds*:

(d) $* \neq \Gamma = \partial\omega$,

(e) $* \neq \Gamma \neq \partial\omega$, $A(x)\nu(x) \equiv \beta(x)\gamma(x)$, $x \in S \neq *$, *where $\beta(x)$ is a scalar function.*

Then the assumptions of Theorem 5.1 *hold for the problem* $T'[0]w = y$ *with* $T'[0]$ *from* (7.7). *The functional* $l \in Y_c^*$ *and the vector* $e \in Y$ *are given by the equalities*

$$l(\{f;\chi\}) = \int_\omega e^\psi f\,dx - \int_S e^\psi (A\nu,\gamma)_n \chi\,dS,$$

$$e = \{f_0;\chi_0\} : \int_\omega e^\psi f_0\,dx - \int_S e^\psi (A\nu,\gamma)_n \chi_0\,dS = 1,$$

$$f_0 \in C^{k-1,\varepsilon}(\overline{\omega}), \qquad \chi_0 \in C^{k,\varepsilon}(\partial\omega), \quad \operatorname{supp}\chi_0 \subset S,$$

where ψ *is the function from the decomposition* (1.5) (*Chapter* 1) *of the coefficient* $a(x)$ *defined in* (7.4).

PROOF. We write the equation $T'[0]w = y$, $y \in Y$, $w \in X$, as the equivalent problem (7.13). By the condition (d), the extended function $\widehat{\chi}$ coincides with χ and, by the condition (e), $\widehat{\chi}(x) = 0$ for $x \in \Gamma$ (see the definition of the space Z). Since $\widehat{\chi}(x)$ is uniquely constructed, the problems (7.13) and (7.15) are equivalent. By Theorem 5.2 from Chapter 1, the equality (5.2) from Chapter 1 is a criterion for solvability of (7.15). In the case under consideration, (5.2) is equivalent to the relation

$$\int_\omega e^\psi f\,dx - \int_S e^\psi (A\nu,\gamma)_n \chi\,dS = 0,$$

which leads to formulas for l and e. The boundedness of l in any Y_c is obvious. □

Now we are in a position to obtain information about the structure of the set of all solutions to the problem (7.9) if the conditions (a)–(c) of Theorem 7.1 fail.

THEOREM 7.2. *Let* $r \geq 3$. *Suppose that the problem* $L^*q = 0$, $q \in C^1(\overline{\omega})$, *has a nonzero solution and one of the following conditions holds*:

(d) $* \neq \Gamma = \partial\omega$,

(e) $* \neq \Gamma \neq \partial\omega$, $A(x)\nu(x) \equiv \beta(x)\gamma(x)$, $x \in S \neq *$, *where* $\beta(x)$ *is some scalar function.*

Consider the quadratic form

$$\theta[u,u] = \int_\omega e^\psi F''[z](u,u)\,dx - \int_S e^\psi (A\nu,\gamma)_n \varphi''[z](u,u)\,dS, \quad u \in N. \tag{7.17}$$

1. *If the quadratic form* $\theta[u,u]$ *is indefinite in* N, *then* $y \equiv z$ *is a nonisolated solution to the problem* (7.1), (7.2) *for any natural number* k.

2. *If for* $\langle\cdot\rangle = \|\cdot\|_2$ *or* $\langle\cdot\rangle = \|\cdot\|_{1,2}$, *the following estimate holds*:

$$|\theta[u,u]| \geq C\langle u\rangle^2, \quad C > 0, \quad u \in N, \tag{7.18}$$

then $y \equiv z$ *is an isolated solution of class* $C^{k,\varepsilon}(\overline{\omega}, \mathbb{R}^n)$ *to the problem* (7.1), (7.2) *for* $k \geq k_0$, *where*

$$k_0 = \begin{cases} 3(2+m/2) & \text{if } \langle\cdot\rangle = \|\cdot\|_2, \\ k_0 = 4 + 3m/2 & \text{if } \langle\cdot\rangle = \|\cdot\|_{1,2}. \end{cases}$$

PROOF. By Lemma 7.1, we can apply Theorem 5.1 to the problem (7.9). It remains to note that the quadratic form (7.17) coincides with the second differential of the left-hand side of the bifurcation equation, and take into account the multiplicative inequalities (cf. the Appendix). □

On the sets (2.1) and (2.8), we define the functions $F(p,q,x,\lambda)$, $p,q \in V_\rho$, $x \in \overline{\omega}$, and $\varphi(q,x,\lambda)$, $q \in V'_\rho$, $x \in \overline{\omega}$, depending on an additional parameter λ. Assume that these functions satisfy the conditions (2.2) and (2.9), in which the differential D^γ_λ is included, the inequality in (2.2) is replaced by $|\alpha|+|\beta|+|\gamma| \le r+1$ and the inequality in (2.9) is replaced by $|\beta|+|\gamma| \le r+1$. In the set (2.4), we state the problem

$$\begin{aligned} F[y,\lambda] &\equiv F(\dot{y}(x), y(x), x, \lambda) = 0, \quad x \in \omega, \\ \varphi[y,\lambda] &\equiv \varphi(y(x), x, \lambda) = 0, \quad x \in S, \\ y(x) - z(x) &= 0, \quad x \in \Gamma u \overline{x}. \end{aligned} \tag{7.19}$$

Assume that $y(x) = z(x)$ satisfies (7.19) for all λ of small modulus. Our purpose is to find sufficient conditions under which $\lambda = 0$ is a bifurcation point for the problem (7.19).

If $\lambda = 0$, then Theorems 7.1 and 7.2 can be applied to the problem (7.19). If for $\lambda = 0$ the assumptions of Theorem 7.1 hold, then $y = z$ is a nonisolated solution and $\lambda = 0$ is a bifurcation point. If for $\lambda = 0$ the assumptions of Theorem 7.2 hold and the quadratic form (7.17) is indefinite, then $y \equiv z$ is again a nonisolated solution and $\lambda = 0$ is a bifurcation point. In the listed cases, the parameter λ does not appear. If the assumptions of Theorem 7.2 for $\lambda = 0$ and the inequality (7.18) hold, then $y \equiv z$ is an isolated solution of class $C^{k,\varepsilon}(\overline{\omega}, \mathbb{R}^n)$ for sufficiently large k. In this case, bifurcation is possible but only at the expense of the parameter λ.

THEOREM 7.3. *Let $r \ge 3$ and let the operator L be found from F for $\lambda = 0$. Suppose that the problem $L^* q = 0$, $q \in C^1(\overline{\omega})$, has a nonzero solution and one of the conditions* (d) *or* (e) *of Theorem* 7.2 *holds. If for some $u_0 \in N$ in* (7.8)

$$\int\limits_\omega e^\psi F_{y\lambda}[z,0] u_0 \, dx - \int\limits_S e^\psi (A\nu, \gamma)_n \varphi_{y\lambda}[z,0] u_0 \, dS \ne 0, \tag{7.20}$$

then $\lambda = 0$ is a bifurcation point.

PROOF. By Lemma 7.1 we can apply Theorem 6.1, which means that (7.20) is a sufficient condition for bifurcation. □

§8. Rigid conditions

As was proved in §7 (cf. Theorem 7.1, the case $\Gamma, S, \overline{x} = *$), the set of all solutions to the problem (7.1), (7.2) has surface structure at a point z. We call a condition (on a set) rigid if a set satisfying this condition must consist of a single point $z(x)$. Using this notion, we can interpret Theorems 7.1 and 7.2 as sufficient conditions for the rigidity (or nonrigidity) of (7.2) on the boundary $\partial\omega$ and at the points $x_j \in \omega$, $j = 1, \dots, l$. In particular, these theorems mean that the conditions imposed at points $x_j \in \overline{x}$ cannot be rigid.

In this section, we study the rigidity of various conditions different from (7.2). For every class of conditions the notion of rigidity has specific features.

8.1. The rigidity of the condition $f(y,x)=0$. We suppose that a function $F(p,q,x)$ satisfies (2.2) and its differential satisfies (7.4) and (7.5). For $f(q,x)$ the conditions (2.9) are assumed to hold with the same r and k as in the case of $F(p,q,x)$. We consider the problem

$$F[y] \equiv F\big(\dot{y}(x), y(x), x\big) = 0, \quad x \in \omega, \\ y(x) - z(x) = 0, \quad x \in \partial\omega, \quad y \in U_\rho, \tag{8.1}$$

where U_ρ is defined in (2.4). As above, $z(x)$ is assumed to satisfy (8.1). We impose additional conditions on the function f and the mapping z:

$$l(x) \notin RA(x), \quad |l(x)| = 1, \quad x \in \omega, \quad l(x) = f_q(z(x), x), \\ A_{ij}(x) = F_{p^i_j}\big(\dot{z}(x), z(x), x\big). \tag{8.2}$$

Under these assumptions, the following theorem holds.

THEOREM 8.1. (a) *Let $r \geq 1$ and let (8.2) hold. Suppose that $a(x)$ from (7.4) does not admit the representation*

$$a(x) = A(x)\nabla\psi(x) + \alpha(x)l(x) \tag{8.3}$$

for any functions $\psi \in C^{k+1,\varepsilon}(\overline{\omega})$ and $\alpha \in C^{k,\varepsilon}(\overline{\omega})$. Then, for any k, $y \equiv z$ is an isolated solution of class $C^{k,\varepsilon}(\overline{\omega}, \mathbb{R}^n)$ to the problem

$$F[y] = 0, \quad f[y] \equiv f\big(y(x), x\big) = 0, \quad x \in \omega, \\ y(x) - z(x) = 0, \quad x \in \partial\omega, \quad y \in U_\rho, \tag{8.4}$$

(b) *Let $r \geq 3$, and let the condition (8.2) hold. Suppose that $a(x)$ from (7.4) is represented in the form (8.3) with some $\psi \in C^{k+1,\varepsilon}(\overline{\omega})$ and $0 \not\equiv \alpha \in C^{k,\varepsilon}(\overline{\omega})$. If the quadratic form*

$$\theta[u,u] = \int_\omega e^\psi \big[F''[z](u,u) - \alpha f_{qq}[z](u,u)\big]\, dx, \\ u \in N_l = \big\{u \in C^{k,\varepsilon}(\overline{\omega}, \mathbb{R}^n) : u|_{\partial\omega} = 0, (u,l)_n = 0, \quad Lu = 0 \text{ at } x \in \overline{\omega}\big\}, \tag{8.5}$$

is indefinite in N_l, then $y \equiv z$ is a nonisolated solution of class $C^{k,\varepsilon}(\overline{\omega}, \mathbb{R}^n)$ to the problem (8.4) for any k. If $\theta[u,u]$ satisfies the estimate

$$|\theta[u,u]| \geq C\langle u\rangle^2, \quad C > 0, \quad u \in N_l, \tag{8.6}$$

where $\langle\cdot\rangle = \|\cdot\|_2$ or $\langle\cdot\rangle = \|\cdot\|_{1,2}$, then for $k \geq k_0$ the function $y \equiv z$ is an isolated solution of class $C^{k,\varepsilon}(\overline{\omega}, \mathbb{R}^n)$ to the problem (8.4). The number k_0 is the same as in Theorem 7.2.

PROOF. (a) Since (8.3) fails, the representation $a(x) = A(x)\nabla\psi(x)$ cannot exist. By Theorem 2.1 in Chapter 1, the problem $L^*q = 0$, $q \in C^1(\overline{\omega})$, has only the zero solution. By the first claim in Theorem 7.1, any solution to the problem (8.1) admits the representation (7.10). Substituting (7.10) in the second equality in (8.4), we obtain

$$F[u] \equiv f\big(z + u + R[u], x\big) = 0, \\ u \in X = \big\{u \in C^{k,\varepsilon}(\overline{\omega}) : u|_{\partial\omega} = 0, Lu = 0\big\}, \quad |u|_{k,\varepsilon} < \delta. \tag{8.7}$$

The set of all solutions to the system (8.4) consists of exactly those y from (7.10) for which u is a solution to the problem (8.7). The mapping $F : B_\delta(0) \subset X \to C^{k,\varepsilon}(\overline{\omega})$ has r continuous derivatives and satisfies the equalities

$$F[0] = 0, \quad F'[0]u = f'[z]u = (l, u)_n,$$
$$F''[0](u,u) = f''[z](u,u) + (l, R''[0](u,u))_n.$$

Since (8.3) is impossible, the assumptions of Theorem 4.1 hold (cf. Remark 8.1 in Chapter 1) for $F'[0]w = y$, $w \in X$, $y \in C^{k,\varepsilon}(\overline{\omega})$. Therefore, the set of all solutions to the equation (8.7) has surface structure at the point $u = 0$, whence the solution $y \equiv z$ to the problem (8.4) is nonisolated.

(b) Let (8.3) hold. By Remark 8.1 in Chapter 1, the assumptions of Theorem 5.1 hold for the problem $F'[0]w = y$, $w \in X$, $y \in C^{k,\varepsilon}(\overline{\omega})$; moreover, the functional l is given by the equality

$$l(y) = \int\limits_\omega e^\psi \alpha y \, dx$$

and the closures of X and $C^{k,\varepsilon}(\overline{\omega})$ in the $C^{s,\varepsilon}(\overline{\omega})$-norms, $1 \le s \le k$, are taken for X_c and Y_c. As in (a), we can establish (7.10) for a solution to the problem (8.1). The bifurcation equation (5.4) for $F[u]$ from (8.7) takes the form

$$\Psi[u] = \int\limits_\omega e^\psi \alpha f(z + u + R[u], x) \, dx = 0, \qquad u \in N_l, \quad |u|_{k,\varepsilon} < \delta,$$

and the second differential of Ψ satisfies

$$\Psi''[0](u,u) = \int\limits_\omega e^\psi \alpha \big[f''[z](u,u) + (l, R''[0](u,u))_n\big] \, dx. \tag{8.8}$$

From (8.3) we obtain $e^\psi[a - A\nabla\psi] = \big[e^\psi a - A\nabla e^\psi\big] = e^\psi \alpha l$. Multiplying (in the sense of the inner product in $\mathbb{R}^n$) both sides of the last equality by $w = R''[0](u,u)$, we arrive at the relation $e^\psi \alpha (l, w)_n = e^\psi (a, w)_n - \big(\nabla e^\psi, A^* w\big)_n$. Integrating the latter over ω and then integrating by parts and using (4.10), we find that

$$\int\limits_\omega e^\psi \alpha (l, w)_n \, dx = \int\limits_\omega e^\psi L w \, dx = -\int\limits_\omega e^\psi F''[z](u, u,) \, dx.$$

Using this relation, we can eliminate the last term on the right-hand side of (8.8). Then

$$\Psi''[0](u,u) = \int\limits_\omega e^\psi [\alpha f''[z](u,u) - F''[z](u,u)] \, dx.$$

To complete the proof, it remains to apply Theorem 5.1 and the multiplicative inequalities from the Appendix. □

REMARK 8.1. If (8.3) is impossible or $\alpha(x) \not\equiv 0$, then the set of all solutions to the problem (8.1) has surface structure at a point z. Therefore, $y \equiv z$ is a nonisolated solution to (8.1). Consequently, Theorem 8.1 yields sufficient conditions for the rigidity (or nonrigidity) of the condition $f(y(x), x) = 0$. If (8.3) holds with $\alpha(x) \equiv 0$, then formula (7.10) for a solution to the problem (8.1) becomes false.

8.2. F-rigid surfaces. We consider the surface $\Sigma = \{y \in \mathbb{R}^n : f(y) = 0\}$, where $n = m+1$, and fix some parametrization of Σ, i.e., a mapping $z \in C^{k,\varepsilon}(\overline{\omega}, \mathbb{R}^n)$ ($\omega \subset \mathbb{R}^m$ is a bounded domain with boundary $\partial\omega \in C^{k+1,\varepsilon}$) such that $f(z(x)) = 0$. We assume that the function f satisfies (2.9) with some r and k, the function F satisfies (2.2) with the same r and k, and the differential of F satisfies (7.4), (7.5). Let $F(\dot{z}(x), z(x), x) = 0$, $x \in \omega$, and

$$l(x) \notin RA(x), \quad |l(x)| = 1, \quad x \in \omega, \quad l(x) = f_q(z(x)). \tag{8.9}$$

If a mapping $y \in C^{k,\varepsilon}(\overline{\omega}, \mathbb{R}^n)$, $y \not\equiv z$, belongs to a sufficiently small $C^{k,\varepsilon}(\overline{\omega}, \mathbb{R}^n)$-neighborhood of z and coincides with z on $\partial\omega$, it is called a small deformation of the surface Σ.

DEFINITION 8.1. A surface Σ is called

F-rigid in $C^{k,\varepsilon}(\overline{\omega}, \mathbb{R}^n)$ if it does not admit small deformations satisfying the equation $F(\dot{y}(x), y(x), x) = 0$,

internally F-rigid in $C^{k,\varepsilon}(\overline{\omega}, \mathbb{R}^n)$ if it does not admit small deformations into itself satisfying the equation $F(\dot{y}(x), y(x), x) = 0$,

geometrically F-rigid in $C^{k,\varepsilon}(\overline{\omega}, \mathbb{R}^n)$ if it does not admit small deformations satisfying the conditions $y(w) \neq \Sigma$ and $F(\dot{y}(x), y(x), x) = 0$.

For $* \neq \Gamma = \partial\omega$, $\overline{x} = *$ Theorem 7.2 yields sufficient conditions for F-rigidity of Σ and Theorem 8.1 provides sufficient conditions for the internal F-rigidity of Σ. Now we turn to the notion of geometrical F-rigidity.

THEOREM 8.2. *Let the problem $L^*q = 0$, $q \in C^1(\overline{\omega})$, with the operator L given by* (7.4) *have only the zero solution. Then the surface Σ is not geometrically F-rigid for any r and k.*

PROOF. By Theorem 7.1, for all small deformations of Σ satisfying the equation $F(\dot{y}(x), y(x), x) = 0$ formula (7.10) is valid, where N is taken from (7.8) for $* \neq \Gamma = \partial\omega$, $\overline{x} = *$. To determine $u \in N$, we consider the problem

$$\frac{d}{dt} f(z + tu + R[tu])|_{t=0} \equiv (l, u)_n = \varphi \in C^{k,\varepsilon}(\overline{\omega}), \quad \varphi\big|_{\partial\omega} = 0.$$

Remark 8.1 from Chapter 1 shows that this problem is solvable either for all or for some $\varphi(x) \not\equiv 0$. If $t \neq 0$, then $y(x,t) = z(x) + tu(x) + R[tu](x)$ is a small deformation of Σ and satisfies the equation $F(\dot{y}, y, x) = 0$. Since $f(y(x,t)) \not\equiv 0$ for $t \neq 0$ small enough, for such t the set $y(\omega, t)$ does not coincide with Σ. □

8.3. The rigidity of the condition $H(\dot{y}, y, x) = 0$. Let the assumptions of Theorem 7.1 with $r \geq 3$ and $S = *$ be satisfied by $F(p, q, x)$ and $z(x)$. Then the set of all solutions to the problem (7.1), (7.2) has surface structure at the point z, i.e., (7.10) holds with (7.11) for $\varkappa = 0, 1, 2$ and (7.12). Let $H(p, q, x)$ be defined in the same neighborhood (2.1) as F; moreover, suppose the derivatives $D^\alpha_{p,q} H(p, q, x)$, $|\alpha| \leq 3$, exist and are continuous for $p, q \in V_\rho$, $x \in \overline{\omega}$. Then $H[y] = H(\dot{y}, y, x)$ is defined in the set (2.4). For the derivatives of $H[y]$ we use the notation (2.6). We assume that

$$H[z] = 0, \quad H'[z]h = F'[z]h = Lh. \tag{8.10}$$

THEOREM 8.3. *Under the above assumptions, for some* $b \geq 1$ *and all* $u \in J^{k,\varepsilon}(\Gamma, *, \overline{x})$ *the inequality*

$$\int_\omega \left|F''[z](u,u) - H''[z](u,u)\right|^b dx \geq C\langle u\rangle^{2b}, \quad C > 0, \tag{8.11}$$

holds, where $\langle\cdot\rangle = \|\cdot\|_{2b}$ *or* $\langle\cdot\rangle = \|\cdot\|_{1,2b}$ *and* $J^{k,\varepsilon}(\Gamma, *, \overline{x})$ *is defined in* (7.8). *Then* $y(x) \equiv z(x)$ *is an isolated solution to the problem*

$$H[y] = F[y] = 0, \quad x \in \overline{\omega}, \quad y(x) - z(x) = 0, \quad x \in \Gamma \cup \overline{x}, \tag{8.12}$$

in $C^{k,\varepsilon}(\overline{\omega}, \mathbb{R}^n)$ *for* $k \geq k_0(b)$, *where*

$$k_0(b) = 3(2 + m/2b) \text{ if } \langle\cdot\rangle = \|\cdot\|_{2b},$$
$$k_0(b) = 4 + 3m/2b \text{ if } \langle\cdot\rangle = \|\cdot\|_{1,2b}.$$

PROOF. By the above assumptions on H, for $y = z + h$ we have

$$H[y] = Lh + 2^{-1}H''[z](h,h) + O(|h|_1^3) \text{ as } |h|_1 \to 0. \tag{8.13}$$

A solution $y(x)$ to the problem (7.1), (7.2) with $S = *$ is represented as $h = u + R[u]$ (cf. (7.10)). From (7.11) we derive the estimates

$$\big|R[u]\big|_{1,\varepsilon} \leq C|u|_{1,\varepsilon}^2, \quad \big|R[u] - 2^{-1}R''[0](u,u)\big|_{1,\varepsilon} \leq C|u|_{1,\varepsilon}^3. \tag{8.14}$$

Since the C^1-norm is majorized by the $C^{1,\varepsilon}$-norm, (8.13) implies

$$H[y] = 2^{-1}\big[H''[z](u,u) + LR''[0](u,u)\big] + O\big(|u|_{1,\varepsilon}^3\big) \text{ as } |u|_{1,\varepsilon} \to 0. \tag{8.15}$$

Using (7.12), from (8.15) we obtain

$$H[y] = 2^{-1}\big[H''[z](u,u) - F''[z](u,u)\big] + O\big(|u|_{1,\varepsilon}^3\big) \text{ as } |u|_{1,\varepsilon} \to 0. \tag{8.16}$$

After raising the moduli of both sides of (8.16) to the power b and integrating the result over ω, in view of the estimate $\big||\alpha| - |\beta|\big|^b \geq 2^{-1}|\alpha|^b - c|\beta|^b$ and the assumption (8.11), we find that

$$\int_\omega |H[y]|^b\, dx \geq C_1\langle u\rangle^{2b} - C_2|u|_{1,\varepsilon}^3 b, \qquad u \in J^{k,\varepsilon}(\Gamma, *, \overline{x}), \quad |u|_{k,\varepsilon} \leq \delta. \tag{8.17}$$

By the multiplicative inequality (cf. the Appendix) $|u|_{1,\varepsilon}^3 \leq C\langle u\rangle^2 |u|_{k,\varepsilon}$, $k \geq k_0(b)$, for δ small enough the inequality (8.17) implies

$$\int_\omega |H[y]|^b\, dx \geq C'\langle u\rangle^{2b}\big[1 - C''|u|_{k,\varepsilon}\big] \geq C\langle u\rangle^{2b}. \tag{8.18}$$

Multiplying (8.18) by $u_{k,\varepsilon}^b$ and taking into account the boundedness of the factor and the multiplicative inequality, we find that

$$\int_\omega |H[y]|^b\, dx \geq C|u|_{1,\varepsilon}^{3b}.$$

By (7.11), $|y - z|_{1,\varepsilon} \leq C|u|_{1,\varepsilon}$. Therefore,

$$\int_\omega |H[y]|^b\, dx \geq C|y - z|_{1,\varepsilon}^{3b}, \quad C > 0. \tag{8.19}$$

This proves the theorem. □

Theorem 8.3 gives sufficient conditions for rigidity of the condition $H(\dot{y}, y, x) = 0$ imposed on solutions to the problem (7.1), (7.2) in the case $S = *$.

8.4. The rigidity of permutation. Let a function $F(p, q, x)$ and a mapping $z(x)$ satisfy (2.2) with some natural numbers r and k, and also (7.4), (7.5). Let

$$f \in C^{r+k+2}(\mathbb{R}^l, \mathbb{R}^n), \quad \zeta \in C^{k,\varepsilon}(\overline{\omega}, \mathbb{R}^l), \quad z(x) = f(\zeta(x)).$$

The Jacobi matrix $f'(q)$ composed of the derivatives with respect to q acts from $\mathbb{R}^l$ into $\mathbb{R}^n$. Consequently, $f'^*(q)$ maps $\mathbb{R}^n$ into $\mathbb{R}^l$. We require that

$$RA(x) \cap Nf'^*(\zeta(x)) = \{0\} \text{ for all } x \in \overline{\omega} \tag{8.20}$$

in order to describe the set of solutions $y \in C^{k,\varepsilon}(\overline{\omega}, \mathbb{R}^n)$ to the problem

$$F[y] = 0, \quad (y - z)\big|_{\partial\omega} = 0, \tag{8.21}$$

which can be represented as

$$\begin{gathered} y(x) = f\big(\zeta(x) + h(x)\big), \\ h \in C^{k,\varepsilon}(\overline{\omega}, \mathbb{R}^l), \quad h\big|_{\partial\omega} = 0, \quad |h|_{k,\varepsilon} < \delta, \end{gathered} \tag{8.22}$$

where δ is sufficiently small. Under the above assumptions, the following theorem holds.

THEOREM 8.4. (a) *Let $r \geq 1$, and let the coefficient $a(x)$ of the operator L not admit the representation*

$$a(x) = A(x)\nabla\psi(x) + \mu(x) \tag{8.23}$$

for any functions $\psi \in C^{k+1,\varepsilon}(\overline{\omega})$, $\mu \in C^{k,\varepsilon}(\overline{\omega}, \mathbb{R}^n)$, $\mu \in Nf'^*(\zeta(x))$. Then the set of all solutions to the problem (8.21), (8.22) has surface structure at $z \in C^{k,\varepsilon}(\overline{\omega}, \mathbb{R}^n)$ for any k.

(b) *Let $r \geq 3$, and let the coefficient $a(x)$ admit the representation (8.23). If the quadratic form*

$$\begin{gathered} \theta[\eta, \eta] = \int\limits_{\omega} e^{\psi}\left[F''[z](f'(\zeta)\eta, f'(\zeta)\eta) + (\mu, f_{q_i q_j}(\zeta)\eta^i\eta^j)_n\right] dx, \\ \eta \in C^{k,\varepsilon}(\overline{\omega}, \mathbb{R}^n), \quad \eta\big|_{\partial\omega} = 0, \quad Lf'(\zeta)\eta = 0, \end{gathered} \tag{8.24}$$

is indefinite, then $z(x)$ is a nonisolated solution of class $C^{k,\varepsilon}(\overline{\omega}, \mathbb{R}^n)$ to the problem (8.21), (8.22) for any k. If for all η the inequality $|\theta[\eta, \eta]| \geq C\langle\eta\rangle^2$ holds, where $\langle\cdot\rangle = \|\cdot\|_2$ or $\langle\cdot\rangle = \|\cdot\|_{1,2}$, then $z(x)$ is the isolated solution of class $C^{k,\varepsilon}(\overline{\omega}, \mathbb{R}^n)$, $k \geq k_0$, which was defined in Theorem 7.2.

PROOF. The problem (8.21), (8.22) can be written as follows:

$$\begin{gathered} T[h] = F[f(\zeta + h)] = 0, \\ h \in C^{k,\varepsilon}(\overline{\omega}, \mathbb{R}^l), \quad h\big|_{\partial\omega} = 0, \quad |h|_{k,\varepsilon} < \delta. \end{gathered} \tag{8.25}$$

The mapping T, regarded as a mapping from $C^{s,\varepsilon}(\overline{\omega}, \mathbb{R}^l)$ into $C^{s-1,\varepsilon}(\overline{\omega})$ for all $s = 1, \ldots, k$, has r continuous derivatives since it is the composition of mappings of the required smoothness (cf. Lemmas 2.1 and 2.2). An easy computation leads to the equalities

$$T'[0]h = Lf'(\zeta)h, \tag{8.26}$$
$$T''[0](h,h) = F''[z](f'(\zeta)h, f'(\zeta)h) + Lf_{q_iq_j}(\zeta)h^ih^j$$

for $h \in C^{s,\varepsilon}(\overline{\omega}, \mathbb{R}^l)$, $h|_{\partial\omega} = 0$. The composition $Lf'(\zeta)$ can be written in the canonical form (1.4) of Chapter 1, with the coefficients

$$A_f(x) = f'^*(\zeta(x))A(x), \quad a_f(x) = f'^*(\zeta(x))a(x), \tag{8.27}$$

where $A_f(x)$ is a matrix-valued function from $\mathbb{R}^m$ into $\mathbb{R}^l$ and $a_f(x)$ is an l-dimensional vector-valued function. It is obvious that $A_f(x)$ and $a_f(x)$ belong to $C^{k,\varepsilon}(\overline{\omega})$. Since the symmetric matrix $G(x)$ is nonsingular and the symmetric $m\times m$ matrix $G_f(x) = A_f^*(x)A_f(x)$ is nonnegative definite, the condition $\det G_f(x) \neq 0$ in $\overline{\omega}$ is equivalent to the condition (8.20). Thus, the operator $L_f(\cdot) \equiv Lf'(\zeta)(\cdot)$ satisfies all the requirements (1.1), (1.2) from Chapter 1. The problem $L_f^*q = 0$, $q \in C^1(\overline{\omega})$, has a nonzero solution if and only if $a_f = A_f\nabla\psi$ for some $\psi \in C^{k+1,\varepsilon}(\overline{\omega})$. The last equality can be rewritten in the form $f'^*(\zeta)(a - A\nabla\psi) = 0$, which is equivalent to (8.23). If the representation (8.23) exists, then for a nonzero solution we can take any function that is proportional to $q(x) = e^\psi$, and $L^*e^\psi = e^\psi\mu$. Therefore, for $h \in C^{k,\varepsilon}(\overline{\omega}, \mathbb{R}^l)$, $h|_{\partial\omega} = 0$, we have

$$\int\limits_\omega e^\psi Lf_{q_iq_j}(\zeta)h^ih^j\,dx = \int\limits_\omega e^\psi\left(\mu, f_{q_iq_j}(\zeta)h^ih^j\right)_n dx$$

and the quadratic form (8.24) coincides with the second differential of the mapping $T[h]$ in the bifurcation equation.

We complete the proof in the same way as the proof of Theorem 7.2. □

REMARK 8.2. If the condition (8.23) with $\mu \not\equiv 0$ holds, then the problem $L^*q = 0$, $q \in C^1(\overline{\omega})$, has only the zero solution. Indeed, the inclusion $\mu \in Nf'^*(z(x))$ and the condition (8.20) imply that in the decomposition of $\mu(x)$ into the sum of two orthogonal terms, $b \in NA^*(x)$ does not identically vanish. Hence the required assertion follows from Theorem 2.1 of Chapter 1. To the problem (8.21) we can apply Theorem 7.1 which asserts that the set of all solutions to the problem (8.21) has surface structure at the point z. If we seek a solution to the problem (8.21) in the form (8.22), then $z(x)$ may be an isolated solution, depending on the behavior of the quadratic form (8.24). Thus, Theorem 8.4 yields sufficient conditions for the rigidity (or nonrigidity) of the condition (8.22).

8.5. The rigidity of the change of variables. We consider those solutions $y(x)$ to the equation (7.1) that admit the form

$$y(x) = z(\xi(x)), \tag{8.28}$$

where the diffeomorphism $\xi(x)$ of ω belongs to a sufficiently small $C^{k,\varepsilon}(\overline{\omega}, \mathbb{R}^m)$-neighborhood of the identity mapping and satisfies the conditions

$$\xi(x) - x = 0, \quad x \in \Gamma \cup \overline{x}, \quad \varphi(\xi(x)) = 0, \quad x \in S, \tag{8.29}$$

where $\varphi(\xi) \equiv d(\xi)$ is the oriented distance between ξ and $\partial\omega$ (cf. the Appendix). If the mapping $\xi(x)$ is sufficiently close to the identity mapping in $C^1(\overline{\omega}, \mathbb{R}^m)$, then it is automatically a diffeomorphism between the domains ω and $\xi(\omega)$; moreover, if $\Gamma \neq *$ and $S \neq *$, then ω and $\xi(\omega)$ coincide. Points of $\Gamma \cup \overline{x}$ are fixed, and the set S is invariant under the diffeomorphism $\xi(x)$ provided that $\xi(x)$ satisfies all the conditions (8.29) The equation for $\xi(x)$ is obtained from (7.1) and (8.28) and can be written in the form

$$F_z(\dot{\xi}(x), \xi(x), x) \equiv F(\dot{z}(\xi(x)), z(\xi(x)), x) = 0. \tag{8.30}$$

Thus, the problem is reduced to the description of the set of all mappings $\xi \in C^{k,\varepsilon}(\overline{\omega}, \mathbb{R}^m)$ that satisfy (8.30) and (8.29) and lie in a sufficiently small $C^{k,\varepsilon}(\overline{\omega}, \mathbb{R}^m)$-neighborhood of the identity mapping.

Let $\partial\omega \in C^{r+k+1,\varepsilon}$ and $z \in C^{r+k+1,\varepsilon}(\omega, \mathbb{R}^n)$. Then z can be extended to a mapping that belongs to the same class and is defined in a ball containing $\overline{\omega}$. Let $F(p,q,x)$ satisfy (2.2) with some natural numbers r and k. Then $F_z(p,q,x)$ satisfies (2.2) with the same r and k. Therefore, the mapping $F_z[\xi] \equiv F_z(\dot{\xi}(x), \xi(x), x)$, being a mapping from $C^{s,\varepsilon}(\overline{\omega}, \mathbb{R}^m)$ into $C^{s-1,\varepsilon}(\overline{\omega})$, has r continuous derivatives for every $s = 1, \ldots, k$ and

$$F'_z[x]h = L\dot{z}h, \quad F''_z[x](h,h) = F''[z](\dot{z}h, \dot{z}h) + Lz''(h,h).$$

The operator $L_z(\cdot) \equiv L\dot{z}(\cdot)$ can be written in the canonical form (1.4) in Chapter 1 with coefficients similar to (8.27):

$$A_z(x) = \dot{z}^*(x)A(x), \quad a_z(x) = \dot{z}^*(x)a(x). \tag{8.31}$$

It is obvious that A_z and a_z belong to $C^{k,\varepsilon}(\overline{\omega})$. Let the matrix $G_z(x) = A_z^*(x)A_z(x)$ be nonsingular:

$$\det G_z(x) \neq 0, \quad x \in \overline{\omega}. \tag{8.32}$$

Then L_z satisfies the conditions (1.1), (1.2) in Chapter 1. Under the above assumptions about the smoothness of $\partial\omega$, the function $\varphi(q)$ from (8.29) satisfies the condition (2.9) in some neighborhood of the form $|q-x| \leq \delta$, $x \in \partial\omega$. Furthermore (cf. the Appendix),

$$\begin{aligned} \varphi'[x]h\big|_{\partial\omega} &= (\nu, h)_m\big|_{\partial\omega}, \\ \varphi''[x](h,h)\big|_{\partial\omega} &= -(K(x)h', h')_m\big|_{\partial\omega}, \end{aligned} \tag{8.33}$$

where $K(x)$ is the matrix of principal curvatures of $\partial\omega$ at the point x and h' is the tangent component of the field h.

What we have said above allows us to apply Theorems 7.1 and 7.2 to the problem (8.30), (8.29). Due to these theorems, we are able to know if the condition (8.28) is rigid or not. We must only take into account that the space $N = J^{k,\varepsilon}(\Gamma, S, \overline{x})$ with $\gamma = \nu$ is defined by the operator L_z.

§9. First order problems with invariants of metric tensor

In this section, we consider a number of concrete problems to illustrate the results of §§7 and 8. In these problems, the function F is constructed as an invariant of the metric tensor g_y of the mapping $y : \omega \to \mathbb{R}^m$,

$$g_y(x) = \dot{y}^*(x)\dot{y}(x). \tag{9.1}$$

9.1. Volume preserving mappings. We fix $z \in C^{k,\varepsilon}(\overline{\omega}, \mathbb{R}^m)$, $k \geq 1$, $\varepsilon \in [0,1]$, such that $\det \dot{z}(x) \neq 0$ in $\overline{\omega}$, and define the function $F(\dot{y}, x)$ by the equality

$$F(\dot{y}, x) = \det{}^{1/2} g_y - \det{}^{1/2} g_z. \tag{9.2}$$

In (9.2), the dependence on $\dot{y}$ is included in the first term, and the dependence on x is in the second term. The function (9.2) satisfies (2.2) for any r and some ρ. Therefore, $F[y] = F(\dot{y}, x)$, being a mapping from $C^{s,\varepsilon}(\overline{\omega}, \mathbb{R}^m)$ into $C^{s-1,\varepsilon}(\overline{\omega})$, $s = 1, \dots, k$, is infinitely differentiable.

LEMMA 9.1. *The following equalities hold:*

$$\begin{aligned} F[z] = 0, \quad & F'[z]h = (\det{}^{1/2} g_z)(\operatorname{tr} g_z^{-1} \dot{z}^* \dot{h}), \\ F''[z](h,h) = {} & (\det{}^{1/2} g_z)\big[\operatorname{tr}^2 g_z^{-1} \dot{z}^* \dot{h} + \operatorname{tr} g_z^{-1} \dot{h}^* \dot{h} \\ & - \operatorname{tr}(g_z^{-1/2} \dot{z}^* \dot{h} g_z^{-1/2})^2 - \operatorname{tr}(g_z^{-1/2} \dot{z}^* \dot{h} g_z^{-1/2})(g_z^{-1/2} \dot{z}^* \dot{h} g_z^{-1/2})^* \big]. \end{aligned}$$

PROOF. Let $y = z + h$, $h \in C^{k,\varepsilon}(\overline{\omega}, \mathbb{R}^m)$, and $g_y - g_z = \delta g$. Then $\det^{1/2} g_y = (\det^{1/2} g_z)\det^{1/2}(1 + g_z^{-1}\delta g)$. Since $\det g = \exp(\operatorname{tr}(\ln g))$ holds for any positive definite matrix, we obtain

$$\begin{aligned} & \det{}^{1/2} g_y - \det{}^{1/2} g_z \\ & \quad = (\det{}^{1/2} g_z)\big[2^{-1} \operatorname{tr}(g_z^{-1}\delta g) - 4^{-1} \operatorname{tr}(g_z^{-1}\delta g)^2 + 8^{-1} \operatorname{tr}^2(g_z^{-1}\delta g)\big] + \beta_1, \end{aligned}$$

where

$$|\beta_1|_{s-1,\varepsilon} = O(|\delta g|^3_{s-1,\varepsilon}) \text{ as } |\delta g|_{s-1,\varepsilon} \to 0, \quad s = 1, \dots, k.$$

We note that, under the trace sign, we may cyclically change the position of matrices or replace a matrix by the adjoint matrix. Since $\delta g = \dot{z}^* \dot{h} + \dot{h}^* \dot{z} + \dot{h}^* \dot{h}$, we get

$$\begin{aligned} \operatorname{tr}(g_z^{-1}\delta g) &= 2 \operatorname{tr}(g_z^{-1} \dot{z}^* \dot{h}) + \operatorname{tr}(g_z^{-1} \dot{h}^* \dot{h}), \\ \operatorname{tr}^2(g_z^{-1}\delta g) &= 4 \operatorname{tr}^2(g_z^{-1} \dot{z}^* \dot{h}) + \beta_2, \\ \operatorname{tr}(g_z^{-1}\delta g)^2 &= 2\big[\operatorname{tr}(g_z^{-1/2} \dot{z}^* \dot{h} g_z^{-1/2})^2 \\ & \qquad + \operatorname{tr}(g_z^{-1/2} \dot{z}^* \dot{h} g_z^{-1/2})(g_z^{-1/2} \dot{z}^* \dot{h} g_z^{-1/2})^* \big] + \beta_3, \end{aligned}$$

where

$$|\beta_i|_{s-1,\varepsilon} = O(|h|^3_{s,\varepsilon}) \text{ as } |h|_{s,\varepsilon} \to 0, \quad s = 1, \dots, k, i = 2, 3.$$

Combining the above results, we arrive at the relation

$$\begin{aligned} \det{}^{1/2} g_y - \det{}^{1/2} g_z = {} & (\det{}^{1/2} g_z)\big\{\operatorname{tr}(g_z^{-1} \dot{z}^* \dot{h} + 2^{-1}\big[\operatorname{tr}^2(g_z^{-1} \dot{z}^* \dot{h}) + \operatorname{tr}(g_z^{-1} \dot{h}^* \dot{h}) \\ & - \operatorname{tr}(g_z^{-1/2} \dot{z}^* \dot{h} g_z^{-1/2})^2 - \operatorname{tr}(g_z^{-1/2} \dot{z}^* \dot{h} g_z^{-1/2})(g_z^{-1/2} \dot{z}^* \dot{h} g_z^{-1/2})^* \big]\big\} + \beta_4, \end{aligned}$$

where

$$|\beta_4|_{s-1,\varepsilon} = O(|h|^3_{s,\varepsilon}) \text{ as } |h|_{s,\varepsilon} \to 0, \quad s = 1, \dots, k,$$

which yields the necessary conclusion. □

LEMMA 9.2. *Let $|y - z|_{k,\varepsilon}$ be sufficiently small and let $(y - z)|_{\partial\omega} = 0$. Then*

$$\int_\omega F[y]\, dx = 0.$$

PROOF. It suffices to show that

$$\int_\omega \det \dot y(x)\,dx = \int_\omega \det \dot z(x)\,dx, \quad z, y \in C^1(\overline{\omega}, \mathbb{R}^m), \tag{9.3}$$

provided that $h(x) = y(x) - z(x)$ vanishes on $\partial\omega$. We assume that $z, y \in C^2(\overline{\omega}, \mathbb{R}^m)$ and use the equality $(\det \dot y) dx_1 \wedge \cdots \wedge dx_m = dy_1 \wedge \cdots \wedge dy_m$. Since $y_i = z_i + h_i$, the right-hand side of this equality can be written as follows: $dz_1 \wedge \cdots \wedge dz_m + dh_i \wedge \psi_i$, where the form ψ_i is the exterior product of the forms dh_j and dz_l. The structure of the form ψ_i shows that $d\psi_i = 0$. Therefore, $dh_i \wedge \psi_i = d(h_i\psi_i)$. By the Stokes formula, the integral of $dh_i \wedge \psi_i$ vanishes, and we obtain (9.3). To prove (9.3) for functions of class C^1, we approximate them by functions of class C^2 and pass to the limit. □

LEMMA 9.3. *Let $z \in C^2(\overline{\omega}, \mathbb{R}^m)$, $\det \dot z(x) \neq 0$ in $\overline{\omega}$. Then*

$$\frac{\partial}{\partial x_i}\left[(\det{}^{1/2} g_z) g_z^{ij} \frac{\partial z}{\partial x_j}\right] = 0 \text{ in } \overline{\omega}.$$

PROOF. By Lemma 9.2, the functional

$$J[y] = \int_\omega \det{}^{1/2} g_y\,dx, \qquad y = z + h, \quad h|_{\partial\omega} = 0,$$

is constant for all h with sufficiently small $C^1(\overline{\omega}, \mathbb{R}^n)$-norm. By Lemma 9.1 ($k = 1$, $\varepsilon = 0$), this functional is differentiable. Hence its derivative vanishes at the point z:

$$0 = J'[z]u = \int_\omega (\det{}^{1/2} g_z) \operatorname{tr}(g_z^{-1} \dot z^* \dot u)\,dx, \quad u \in C^1(\overline{\omega}, \mathbb{R}^m), \quad u|_{\partial\omega} = 0.$$

Integrating the last equality by parts and using the arbitrariness of u, we obtain the required equality. □

LEMMA 9.4. *Let $z \in C^{k+1,\varepsilon}(\overline{\omega}, \mathbb{R}^m)$. Then the operator $Lh \equiv F'[z]h$ can be written in the canonical form* (1.4) *from Chapter* 1 *with the coefficients*

$$A(x) = (\det{}^{1/2} g_z) \dot z^{*-1}, \quad a(x) \equiv 0.$$

PROOF. From the formula for $F'[z]h$ we obtain the equality

$$Lh = (\det{}^{1/2} g_z) \operatorname{tr}(g_z^{-1} \dot z^* \dot h) = \frac{\partial}{\partial x_i}\left[(\det{}^{1/2} g_z) g_z^{ik} z^j_{x_k} h^j\right] - h^j \frac{\partial}{\partial x_i}\left[(\det{}^{1/2} g_z) g_z^{ik} z^j_{x_k}\right],$$

which, together with Lemma 9.3, allows us to express A and a. We note that $A_{ij} \in C^{k,\varepsilon}(\overline{\omega})$ and the matrix $G(x) = (\det g_z) g_z^{-1}$ is nonsingular in $\overline{\omega}$. □

We consider the problem

$$\det{}^{1/2} g_y - \det{}^{1/2} g_z = 0, \qquad x \in \overline{\omega}, \quad y - z = 0, \quad x \in \Gamma \cup \overline{x}, \tag{9.4}$$

which consists in describing the set of all $y(x)$ that are sufficiently close to $z(x)$ in the space $C^{k,\varepsilon}(\overline{\omega}, \mathbb{R}^m)$. Such a problem coincides with the problem (7.1), (7.2) for $S = *$ and the function F in (9.2).

THEOREM 9.1. *Let $z \in C^{k+1,\varepsilon}(\overline{\omega}, \mathbb{R}^m)$, $\det g_z(x) \neq 0$ in $\overline{\omega}$. Then the set of all solutions to the problem* (9.4) *has surface structure at the point z. The mapping R in* (7.10) *is infinitely differentiable and satisfies* (7.11).

PROOF. Since the coefficient $a(x)$ of the operator L vanishes, we have $L^* = -A\nabla$ and the problem $L^*q = 0$, $q \in C^1(\overline{\omega})$, has a nonzero solution $q \equiv 1$. Therefore, for $\Gamma \neq \partial\omega$ Theorem 9.1 is a consequence of Theorems 7.1 (the condition (a)) and 7.2.

Let $* \neq \Gamma = \partial\omega$. By Lemma 9.2, $F[y]$ belongs to the subspace Y of the space of functions $f \in C^{k-1,\varepsilon}(\overline{\omega})$ that are orthogonal to 1. For the problem

$$F'[z]h = f, \quad h\big|_{\partial\omega} = 0, \quad h \in C^{k,\varepsilon}(\overline{\omega}, \mathbb{R}^m), \quad f \in Y,$$

all the assumptions of Theorems 4.1 and 4.2 are satisfied for the spaces X_c and Y_c that are the closures of the spaces $X = \{u \in C^{k,\varepsilon}(\overline{\omega}, \mathbb{R}^m) : u(x) = 0, x \in \Gamma \cup \overline{x}\}$ and Y in the $C^{s,\varepsilon}(\overline{\omega}, \mathbb{R}^m)$-norm and $C^{s-1,\varepsilon}(\overline{\omega})$-norm respectively. Therefore, the required assertion for $* \neq \Gamma = \partial\omega$ follows from Theorems 4.1 and 4.2. □

In the case $z(x) \equiv x$, the problem (9.4) is to describe all diffeomorphisms $y(x)$ between ω and $y(\omega)$ that are close to the identity mapping and satisfy the conditions

$$\det y(x) = 1, \quad x \in \overline{\omega}, \quad y(x) - x = 0, \quad x \in \Gamma \cup \overline{x}. \tag{9.5}$$

The conditions (9.5) means that the Euclidean volume is preserved, i.e., $dy(x) = dx$.

9.2. A diffeomorphism of ω preserving the Riemann volume. Let $\partial\omega \in C^{k+r+1,\varepsilon}$ and let a function $\rho \in C^{k+r,\varepsilon}(\overline{\omega})$ be positive. We assume that $\rho(x)$ is extended to a positive function of the same class but the extended function is defined in some ball containing $\overline{\omega}$. We introduce the function

$$F(\dot{y}, y, x) = \rho(y) \det \dot{y} - \rho(x). \tag{9.6}$$

Let $z(x) \equiv x$. Then F satisfies (2.2) for some $\rho > 0$. For $F[y] \equiv F(\dot{y}, y, x)$ we have

$$\begin{gathered} F[x] = 0, \quad F'[x]h \equiv Lh = \operatorname{div} \rho(x)h, \\ F''[x](h,h) = \rho(x)\big[(\operatorname{div} h)^2 - h^i_{x_j} h^j_{x_i}\big] + 2\rho_{x_i}(x) h^i \operatorname{div} h + \rho_{x_i x_j}(x) h^i h^j. \end{gathered} \tag{9.7}$$

The last equality in (9.7) follows from the similar equality in Lemma 9.1 for $z(x) \equiv x$. Let $\varphi(q) \equiv d(q)$ be the oriented distance from q to $\partial\omega$ (cf. the Appendix). Under the above assumption about the smoothness of $\partial\omega$, the function $\varphi(q)$ satisfies (2.9) with $z(x) = x$, and (8.32) is valid for $\varphi'[x]$ and $\varphi''[x]$.

We consider the problem

$$\begin{gathered} \rho(y(x)) \det \dot{y}(x) - \rho(x) = 0, \quad x \in \overline{\omega}, \\ y(x) - x = 0, \quad x \in \Gamma \cup \overline{x}, \varphi(y(x)) = 0, \quad x \in S, \end{gathered} \tag{9.8}$$

which consists in describing the set of solutions of class $C^{k,\varepsilon}(\overline{\omega}, \mathbb{R}^m)$ that belong to a sufficiently small $C^{k,\varepsilon}(\overline{\omega}, \mathbb{R}^m)$-neighborhood of the identity mapping $z(x) \equiv x$. Since such a solution is close to the identity mapping in $C^1(\overline{\omega}, \mathbb{R}^m)$, it is a diffeomorphism between ω and $y(\omega)$. If $\Gamma \neq *$, $S \neq *$, then ω and $y(\omega)$ coincide. The equation (9.8) is equivalent to the equality $\rho(y)dy = \rho(x)dx$, which means that the mapping $y(x)$ preserves the Riemann volume with density ρ.

LEMMA 9.5. *Let $y(x)$ be a diffeomorphism of the domain ω onto itself such that* $\det \dot{y}(x) > 0$, $x \in \overline{\omega}$. *Then*

$$\int_\omega \rho(y(x)) \det \dot{y}(x)\, dx = \int_\omega \rho(x)\, dx.$$

PROOF. Making a change of variables, we obtain

$$\int\limits_\omega \rho(y(x)) \det \dot{y}(x)\, dx = \int\limits_{y(\omega)} \rho(y)\, dy = \int\limits_\omega \rho(x)\, dx. \quad \square$$

LEMMA 9.6. *If $y(x)$ is close to the identity mapping in $C^1(\overline{\omega}, \mathbb{R}^m)$ and is a solution to the problem*

$$\begin{gathered} \rho(y(x)) \det \dot{y}(x) - \rho(x) = 0, \quad x \in \overline{\omega}, \\ \varphi(y(x)) - \chi_0(x) \int\limits_S \rho(x)\varphi(y(x))\, dS = 0, \quad x \in S \neq *, \\ y(x) = x, \quad x \in \Gamma \neq 0*, \quad \Gamma \neq \partial\omega, \end{gathered} \tag{9.9}$$

where $\chi_0(x) \in C(\partial\omega)$ is a fixed function such that

$$\operatorname{supp} \chi_0 \subset S, \quad \chi_0(x) \geq 0, \quad \int\limits_{\partial\omega} \rho(x)\chi_0(x)\, dS = 1,$$

then $\varphi(y(x)) \equiv 0$.

PROOF. Since $y(x)$ is close to the identity mapping in $C^1(\overline{\omega})$, it is a diffeomorphism between ω and $y(\omega)$. Integrating the equation in (9.9) over ω and making a change of variables, we conclude that the Riemann volumes of ω and of $y(\omega)$ are equal, i.e.,

$$\int\limits_{y(\omega)} \rho(x)\, dx = \int\limits_\omega \rho(x)\, dx.$$

The boundary condition on Γ yields the required inclusion $\operatorname{supp} \varphi(y(x)) \subset S$. Assume that the integral of $\rho(x)\varphi(y(x))$ over S does not vanish. Then $\varphi(y(x)) \not\equiv 0$ is nonnegative if the integral is positive, and is nonpositive otherwise. Since $y(x)$ is close to the identity mapping in $C^1(\overline{\omega}, \mathbb{R}^m)$, from the definition of φ we obtain the strict inclusions $y(\omega) \subset \omega$ in the first case and $\omega \subset y(\omega)$ in the second case. These inclusions contradict the fact that the Riemann volumes of ω and of $y(\omega)$ are equal. Therefore, the integral of $\rho(x)\varphi(y(x))$ over S does not vanish. From the boundary condition in (9.9) imposed on S we obtained the required equality. $\square$

THEOREM 9.2. *Let $\partial\omega \in C^{r+k+1,\varepsilon}$ and let $\rho \in C^{r+k,\varepsilon}(\overline{\omega})$ be positive. Then for any natural numbers r and k the set of all solutions to the problem* (9.8) *has surface structure at the point $z(x) \equiv x$. The mapping R in* (7.10) *is r-times continuously differentiable and satisfies* (7.11).

PROOF. The problem (9.8) is the problem (7.1), (7.2) with $z(x) \equiv x$, $\varphi(y) \equiv d(y)$, and F from (9.6). The operator L corresponding to this problem is described in (9.7). For L we have $A \equiv \rho I$, $a \equiv 0$. The problem $L^* q = 0$, $q \in C^1(\overline{\omega})$, has a nonzero solution $q \equiv 1$. If there are pieces of the boundary $\partial\omega$ with no boundary condition imposed on them, the required assertion follows from Theorem 7.1. For $* \neq \Gamma = \partial\omega$ we use Lemma 9.5 and argue as in Theorem 9.1.

Let $* \neq \Gamma \neq \partial\omega$ and $S \neq *$. Since $\gamma(x) \equiv \nu(x)$, the condition (e) in Theorem 7.2 is satisfied. Consequently, the application of this theorem does not lead to the desired result. To prove the required assertion, we write out the splitting scheme (5.1) for the problem (7.1), (7.2) with those F and φ which are of interest for our purpose and use Lemma 7.1:

$$F(\dot{y}, y, x) - \alpha f_0(x) = 0, \quad x \in \overline{\omega}, \quad \varphi(y) - \alpha\chi_0(x) = 0, \quad x \in S,$$
$$y(x) - x = 0, \quad x \in \Gamma \cup \overline{x},$$
$$f_0 \in C^{k-1,\varepsilon}(\overline{\omega}), \quad \chi_0 \in C^{k,\varepsilon}(\partial\omega), \quad \operatorname{supp}\chi_0 \subset S,$$
$$\alpha \equiv \int_\omega F(\dot{y}, y, x)\,dx - \int_S \rho\varphi(y)\,dS = 0, \quad \int_\omega f_0\,dx - \int_S \rho\chi_0\,dS = 1.$$

Take $f_0(x) \equiv 0$ and $\chi_0(x) \leq 0$. By Lemma 9.5, the integral of $F(\dot{y}, y, x)$ vanishes. Hence the last system takes the form

$$\begin{gathered} F(\dot{y}, y, x) = 0, \quad x \in \omega, \\ \varphi(y) + \alpha\chi_0 = 0, \quad x \in S, \quad y(x) - x = 0, \quad x \in \Gamma \cup \overline{x}, \end{gathered} \tag{9.10}$$

$$\alpha \equiv \int_S \rho\varphi(y)\,dS = 0, \quad \int_S \rho\chi_0\,dS = -1. \tag{9.11}$$

By Theorem 5.1, any solution to the problem (9.10) admits a representation (7.10) with an r-times continuously differentiable mapping R satisfying (7.11). By Lemma 9.6, the set of solutions $y(x)$ to the problem (9.10) that belong to a sufficiently small $C^{k,\varepsilon}(\overline{\omega}, \mathbb{R}^m)$-neighborhood of the identity mapping $z(x) \equiv x$ coincides with the set of solutions to the problem (9.8) that are close to the identity mapping $z(x) \equiv x$ in the $C^{k,\varepsilon}(\overline{\omega}, \mathbb{R}^m)$-norm. □

9.3. Equations containing the invariant tr. We introduce the family of functions $F_s(\dot{y})$, $s = 0, 1, \ldots$, by the equalities

$$F_0(\dot{y}) = \det{}^{1/2} g_y - 1, \quad F_s(\dot{y}) = \frac{1}{2s}\left(\operatorname{tr} g_y^s - m\right), \quad s = 1, 2, \ldots. \tag{9.12}$$

The functions (9.12) satisfy (2.2) with $z(x) \equiv x$ for any r and some ρ. A simple computation gets

$$\begin{gathered} F_s'[x]h = \operatorname{div} h, \quad s = 0, 1 \ldots, \\ F_0''[x](h, h) = h^i_{x_i} h^j_{x_j} - h^i_{x_j} h^j_{x_i}, \\ F_s''[x](h, h) = [4(s-1) + 1] h^i_{x_j} h^i_{x_j} + 4(s-1) h^i_{x_j} h^j_{x_i}, \quad s = 1, 2, \ldots. \end{gathered} \tag{9.13}$$

Let M_s, $s = 0, 1, \ldots$, be the set of solutions to the problem

$$F_s[y] = 0, \quad x \in \overline{\omega}, \quad y(x) - x = 0, \quad x \in \Gamma \cup \overline{x}, \tag{9.14}$$

that belong to a sufficiently small $C^{k,\varepsilon}(\overline{\omega}, \mathbb{R}^m)$-neighborhood of the identity mapping $z(x) \equiv x$. Each of the problems (9.14) is a special case of the problem (7.1), (7.2) with F from (9.12) and $S = *$. We assume that $\Gamma \neq *$, but it is possible that Γ is the empty set or coincides with the boundary.

THEOREM 9.3. (a) *For $\Gamma \neq \partial\omega$ every set M_s has surface structure in $C^{k,\varepsilon}(\overline{\omega}, \mathbb{R}^m)$ with a common tangent space N at the point $z(x) \equiv x$ for all s.*

(b) *For $\Gamma = \partial\omega$ the set M_0 has surface structure at the point $z(x) \equiv x$.*

(c) *For $\Gamma = \partial\omega$ the solution $y(x) \equiv x$ is a unique solution to the problem* (9.14) *with $s = 1$ in $C^1(\overline{\omega}, \mathbb{R}^m)$.*

(d) *For $\Gamma = \partial\omega$ and $k \geq 4 + 3m/2$ the set M_s with $s = 2, 3, \ldots$ consists of a single point $y(x) \equiv x$.*

(e) *Let the set $\Gamma \cup \overline{x}$ be such that for $y(x) = Ux + x_0$, where U is an unitary matrix and $x_0 \in \mathbb{R}^m$ is fixed, the equality $y(x) - x = 0$, $x \in \Gamma \cup \overline{x}$, holds only if $U = I$ and $x_0 = 0$. Then for $k > 4$ the set $M_s \cap M_t$, $s, t = 0, 1, \ldots$, $s \neq t$, consists of a single point $y(x) \equiv x$.*

PROOF. (a) For any problem of type (9.14) the operator L coincides with div. Therefore, in the case $\Gamma \neq \partial\omega$, the required assertion follows from Theorem 7.1.

(b) The problem (9.14) with $s = 0$ is a particular case of the problem (9.4). In fact, the assertion is contained in Theorem 9.1.

(c) Let $h(x) = y(x) - x$. Then $F_1[y] = \operatorname{div} h + 2^{-1} \operatorname{tr} \dot{h}^* \dot{h}$. Integrating the equality $F_1[y] = 0$ over ω and taking into account the zero boundary conditions imposed on h, we obtain the relation $\int_\omega h^i_{x_j} h^i_{x_j}\, dx = 0$, which yields $h \equiv 0$.

(d) Since $L^* = -\nabla$, the problem $L^* q = 0$ has a nonzero solution $q \equiv \text{const}$. Therefore, a bifurcation equation arises for the problem (9.14). By (9.13), for $\Psi''[0](u, u)$ and $s \geq 1$ formula (5.5) takes the form

$$\Psi''[0](u,u) = \int_\omega \left[(4(s-1)+1) u^i_{x_j} u^i_{x_j} + 4(s-1) u^i_{x_j} u^j_{x_i}\right] dx,$$

$$u \in C^{k,\varepsilon}(\overline{\omega}, \mathbb{R}^m), \quad u\big|_{\partial\omega} = 0, \quad \operatorname{div} u = 0.$$

Integrating by parts the second term on the right-hand side of the last expression, we find that

$$\Psi''[0](u,u) = 4[(s-1)+1] \int_\omega u^i_{x_j} u^i_{x_j}\, dx \geq C_s \|u\|^2_{1,2}.$$

The required assertion follows from Theorem 7.2.

(e) We regard one of the equations (9.14) with $F = F_s$ as a problem about the description of the set M_s, and the equation with $H = F_t$, $t \neq s$, as an additional condition on M_s. To prove (e), it suffices to show that this condition is rigid. Since $F[x] = H[x] = 0$ and $F'[x] = H'[x]$, we see that such restrictions have been studied in 8.3. Further,

$$F''_s[x](u,u) - F''_t[x](u,u) = \beta_{st}\left(u^i_{x_j} u^i_{x_j} + u^i_{x_j} u^j_{x_i}\right),$$
$$\beta_{st} = 4(s-t), \quad s, t \neq 0, \quad \beta_{s0} = 4(s-1)+1, \quad s \neq 0.$$

Using the relation

$$\begin{aligned} u^i_{x_j}u^i_{x_j} + u^i_{x_j}u^j_{x_i} &= 2^{-1}\big(u^i_{x_j}u^i_{x_j} + 2u^i_{x_j}u^j_{x_i} + u^j_{x_i}u^j_{x_i}\big) \\ &= 2^{-1}\big(u^i_{x_j} + u^j_{x_i}\big)\big(u^i_{x_j} + u^j_{x_i}\big), \end{aligned}$$

we obtain the inequality

$$\int_\omega \big|F_s''[x](u,u) - F_t''[x](u,u)\big|^b\,dx \geq C\int_\omega \sum_{i,j}\big|u^i_{x_j} + u^j_{x_i}\big|^{2b}\,dx,$$

$$u \in C^{k,\varepsilon}(\overline{\omega},\mathbb{R}^m), \quad u(x) = 0, \quad x \in \Gamma \cup \overline{x}, \quad \operatorname{div} u = 0.$$

Let $2b > m$. Since the Korn inequality

$$\int_\omega \sum_{i,j}\big|u^i_{x_j} + u^j_{x_i}\big|^{2b}\,dx \geq C\|u\|^{2b}_{1,2b}, \quad C > 0, \quad u \in W^1_{2b}(\omega,\mathbb{R}^m),$$

$$u(x) = 0, \quad x \in \Gamma \cup \overline{x},$$

holds under the assumptions imposed on $\Gamma \cup \overline{x}$ (cf. [**35**]), we obtain

$$\int_\omega \big|F_s''[x](u,u) - F_t''[x](u,u)\big|^b\,dx \geq C\|u\|^{2b}_{1,2b}.$$

By Theorem 8.3, for $y \in M_s$ and $k \geq k_0(b)$

$$\int_\omega |F_t[y]|^b\,dx \geq C|y - x|^{3b}_{1,\varepsilon}. \tag{9.15}$$

Let $k > 4$. Since $k_0(b) = 4 + 3m/2b$, we have $k \geq k_0(b)$ for sufficiently large b. Thus, $F_t[y] = 0$ is a rigid condition in $C^{k,\varepsilon}(\overline{\omega},\mathbb{R}^m)$ for $k > 4$. □

§10. Variational problems with constraints

10.1. Functionals and constraints. We consider a function $H(p,q,x)$, $p \in \mathbb{R}^{nm}$, $q \in \mathbb{R}^n$, $x \in \Omega \subset \mathbb{R}^m$, such that its derivatives $D^\alpha_{p,q}H(p,q,x)$, $|\alpha| \leq 3$, exist and are continuous in $\mathbb{R}^{nm} \times \mathbb{R}^n \times \overline{\Omega}$, and a function $h(q,x)$, $q \in \mathbb{R}^n$, $x \in \partial\Omega$, such that its derivatives $D^\alpha_q h(q,x)$, $|\alpha| \leq 3$, are continuous in $\mathbb{R}^n \times \partial\Omega$. We introduce the functional

$$J[y] = \int_\Omega H(\dot{y},y,x)\,dx + \int_{\partial\Omega} h(y,x)\,dx \tag{10.1}$$

on the set of mappings $y : \Omega \to \mathbb{R}^n$, and fix $z \in C^{k,\varepsilon}(\overline{\Omega},\mathbb{R}^n)$, where k is a natural number and $\varepsilon \in (0,1]$. Denote by Σ a closed part of the boundary $\partial\Omega$ (the cases $\Sigma = \partial\Omega$ and $\Sigma = \varnothing$ are possible). Let the domain of the functional J consist of mappings $y : \Omega \to \mathbb{R}^n$ satisfying the condition

$$y(x) - z(x) = 0, \quad x \in \Sigma, \tag{10.2}$$

and the constraint

$$F(\dot{y}(x), y(x), x) = 0, \quad x \in \Omega, \tag{10.3}$$

where F is a scalar function.

DEFINITION 10.1. A function $z \in C^{k,\varepsilon}(\overline{\Omega}, \mathbb{R}^n)$ is called a regular point in ω for the functional (10.1) defined on the set of y satisfying (10.2) and (10.3) if the following conditions hold:

(a) ω is a bounded domain with boundary $\partial\omega \in C^{k+1,\varepsilon}$, and the sets $S = \partial\omega \cap (\partial\Omega/\Sigma)$ and $\Gamma = \partial\omega/\overline{S}$ satisfy the requirements of §3,

(b) $F(p,q,x)$ satisfies (2.2) with $r \geq 3$ and some $\rho > 0$, and the differential of $F(p,q,x)$ satisfies (7.5),

(c) for the problem

$$F(\dot{y}, y, x) = 0, \quad x \in \overline{\omega}, \quad y(x) - z(x) = 0, \quad x \in \Gamma, \quad |y - z|_{k,\varepsilon} < \delta, \tag{10.4}$$

the assumptions of Theorem 7.1 with $S = *$, $\overline{x} = *$, $\Gamma \neq *$ are fulfilled and $z(x)$ satisfies the equation $F(\dot{z}(x), z(x), x) = 0$ in Ω.

By Definition 10.1, any solution to the problem (10.4) can be represented in the form (7.10) with δ small enough and an r-times continuously differentiable mapping R satisfying (7.11) and (7.12).

10.2. An increment of a functional with constraint at a regular point. Let $z(x)$ be a regular point for the functional (10.1) with constraint, let $y(x) \equiv z(x)$, $x \in \Omega/\omega$, and let $y(x)$ be a solution to the problem (10.4) in ω. Then $y(x)$ satisfies (10.3) and admits the representation (7.10) in ω.

LEMMA 10.1. *For $u \in J^{k,\varepsilon}(\Gamma, *, *)$, $|u|_{1,\varepsilon} \to 0$, the following relation holds*:

$$\begin{aligned}
J[y] - J[z] = {} & \left[\int_\omega H'[z]u\,dx + \int_S h'[z]u\,dS\right] \\
& + 2^{-1}\left[\int_\omega (H''[z](u,u) + H'[z]R''[0](u,u))\,dx \right. \\
& \left. + \int_S (h''[z](u,u) + h'[z]R''[0](u,u))\,dS\right] + O(|u|_{1,\varepsilon}^3).
\end{aligned}$$

PROOF. For $x \in \Omega/\omega$ the mappings $y(x)$ and $z(x)$ coincide. For $x \in \omega$ the following equalities can be proved as in 8.3:

$$\begin{gathered}
H[y] - H[z] = H'[z]u + 2^{-1}(H''[z](u,u) + H'[z]R''[0](u,u)) + O(|u|_{1,\varepsilon}^3), \\
h[y] - h[z] = h'[z]u + 2^{-1}(h''[z](u,u) + h'[z]R''[0](u,u)) + O(|u|_{1,\varepsilon}^3), \\
u \in J^{k,\varepsilon}(\Gamma, *, *), \quad |u|_{1,\varepsilon} \to 0.
\end{gathered}$$

Integrating the first equality over ω and the second equality over S, we obtain the required assertion. $\square$

For $u \in J^{k,\varepsilon}(\Gamma, *, *)$ we introduce the notation

$$J'[z]u = \int_\omega H'[z]u\,dx + \int_S h'[z]u\,dS,$$

$$J''[z](u,u) = \int_\omega (H''[z](u,u) + H'[z]R''[0](u,u))\,dx + \int_S (h''[z](u,u) + h'[z]R''[0](u,u))\,dS, \tag{10.5}$$

10.3. Critical regular points for a functional with constraint. A regular point $z(x)$ in ω for the functional (10.1) with constraint is said to be critical if $J'[z]u = 0$ for all $u \in J^{k,\varepsilon}(\Gamma, *, *)$.

LEMMA 10.2. *Let $z(x)$ be a critical regular point in ω for the functional* (10.1) *with constraint. For $z \in C^2(\overline{\omega}, \mathbb{R}^n)$ and $H_{p^i_j}(\dot z(x), z(x), x) \in C^1(\overline{\omega})$ there exists a unique function $f \in C^1(\overline{\omega})$ such that*

$$-\frac{d}{dx_i} H_{p^i_j}(\dot z(x), z(x), x) + H_{q^i}(\dot z(x), z(x), x) = (L^* f)^i, \quad x \in \overline{\omega},$$
$$(H_{p^i_j}(\dot z(x), z(x), x) - F_{p^i_j}(\dot z(x), z(x), x) f(x))\nu^j + h_{q^i}(z(x), x)) = 0, \quad x \in S, \tag{10.6}$$

where $j = 1, \dots, n$ and the operator L is defined from F by (7.4).

PROOF. The equality $J'[z]u = 0$ is equivalent to the integral equality

$$0 = \int_\omega (H_{p^i_j} u^i_{x_j} + H_{q^i} u^i)\,dx + \int_S h_{q^i}\,dS, \quad u \in J^{k,\varepsilon}(\Gamma, *, *),$$

where $H = H(\dot z, z, x)$ and $h = h(z, x)$. Integrating it by parts, we obtain

$$0 = \int_\omega \Big(-\frac{d}{dx_j} H_{p^i_j} + H_{q^i} \Big) u^i\,dx + \int_S (H_{p^i_j}\nu^j + h_{q^i}) u^i\,dS. \tag{10.7}$$

For u we take an arbitrary function of $J^{k,\varepsilon}(\partial\omega, \varnothing, *) \subset J^{k,\varepsilon}(\Gamma, *, *)$. Then the integral over S on the right-hand side of (10.7) vanishes and the first equality in (10.6) follows from Theorem 10.1, Chapter 1. Therefore, we can rewrite (10.7) as the following integral equality:

$$\int_\omega (L^* f)^i u^i\,dx + \int_S (H_{p^i_j}\nu^j + h_{q^i}) u^i\,dS = 0.$$

Integrating it by parts and using the equation $Lu = 0$, we find that

$$\int_S \big[(H_{p^i_j} - F_{p^i_j})\nu^j + h_{q^i}\big] u^i\,dS = 0, \quad u \in J^{k,\varepsilon}(\Gamma, *, *). \tag{10.8}$$

The equality (10.8) has meaning only if $S \neq \varnothing$.

If the equation $L^* q = 0$ has only the zero solution, then the trace of u on $\partial\omega$ is a function of class $C^{k,\varepsilon}(\partial\omega, \mathbb{R}^n)$ and its support lies in S. Hence the second equality in (10.6) follows from (10.8).

If the equation $L^*q = 0$ has a nonzero solution, then the function f is determined up to a summand proportional to p^* (cf. Theorem 2.1, Chapter 1). The trace of $u \in J^{k,\varepsilon}(\Gamma, *, *)$ on $\partial\omega$ is a function of class $C^{k,\varepsilon}(\partial\omega, \mathbb{R}^n)$ and its support lies in S; moreover, $\int_S p^*(A\nu, u)_n \, dS = 0$ (cf. Theorem 5.1, Chapter 1). Therefore, (10.8) implies

$$(H_{p^i_j} - F_{p^i_j})\nu^j = \alpha p^* F_{p^i_j}\nu^j, \quad x \in S, \quad i = 1, \dots, n, \tag{10.9}$$

where α is a real constant. If we substitute $f + \alpha p^*$ for f, the first equality in (10.6) remains unchanged whereas (10.9) becomes the second equality in (10.6).

To prove the uniqueness of f, we assume that f_1 and f_2 satisfy (10.6) with the same mapping z. Denoting $f = f_1 - f_2$, for $f \in C^1(\overline{\omega})$ we have $L^*f = 0$ in $\overline{\omega}$ and $fA\nu = 0$ on S. If the equation $L^*f = 0$ has only the zero solution, then $f = 0$. Otherwise (this is possible only if $S \neq \varnothing$, cf. Definition 10.1(c)), $f = \alpha p^*$, and $\alpha = 0$ in view of the boundary condition on S. □

LEMMA 10.3. *If $z \in C^2(\overline{\omega})$, $H_{p^i_j}(\dot{z}(x), z(x), x) \in C^1(\overline{\omega})$, and $f \in C^1(\overline{\omega})$ satisfy* (10.6), *then $z(x)$ is a critical regular point in ω for the functional* (10.1) *with constraint.*

PROOF. The equality $J'[z]u = 0$ for $u \in J^{k,\varepsilon}(\Gamma, *, *)$ is equivalent to (10.7). The latter follows from (10.6). □

10.4. Computation of $J''[z](u, u)$ at a critical regular point.

LEMMA 10.4. *Let $z(x)$ be a critical regular point in ω for the functinal* (10.1) *with constraint. If $z \in C^2(\overline{\omega}, \mathbb{R}^n)$ and $H_{p^i_j}(\dot{z}(x), z(x), x) \in C^1(\overline{\omega})$, then*

$$J''[z](u, u) = \int_\omega \big(H''[z](u, u) - fF''[z](u, u)\big)\, dx + \int_S h''[z](u, u)\, dS, \tag{10.10}$$

where $u \in J^{k,\varepsilon}(\Gamma, *, *)$ and f was introduced in Lemma 10.2.

PROOF. To obtain (10.10), it suffices to compute

$$\begin{aligned} J'[z]R''[0](u, u) = &\int_\omega \big[H_{p^i_j}(R''[0](u, u))^i_{x_j} + H_{q^i}(R''[0](u, u))^i\big]\, dx \\ &+ \int_S h_{q^i}(R''[0](u, u))^i\, dS, \end{aligned} \tag{10.11}$$

where $H = H(\dot{z}, z, x)$ and $h = h(z, x)$. Integrating by parts, we write the right-hand side of (10.11) in the form

$$\int_\omega \Big(-\frac{d}{dx_j}H_{p^i_j} + H_{q^i}\Big)(R''[0](u, u))^i\, dx + \int_S \big(h_{q^i} + H_{p^i_j}\nu^j\big)(R''[0](u, u))^i\, dS.$$

Replacing the expressions in parentheses by their representations from (10.6), we obtain

$$J'[z]R''[0](u, u) = \int_\omega (L^*f, R''[0](u, u))_n\, dx + \int_S fF_{p^i_j}\nu^j(R''[0](u, u))^i\, dS.$$

Integrating by parts once again, we arrive at the relation

$$J'[z]R''[0](u,u) = \int\limits_\omega fLR''[0](u,u)\,dx = -\int\limits_\omega fF''[z](u,u)\,dx. \tag{10.12}$$

Combining (10.11) and (10.12) with the remaining terms in (10.5), we obtain (10.10). □

10.5. Necessary condition for a local minimum of a functional with constraint at a regular point. A regular point $z(x)$ in ω for the functional (10.1) with constraint is said to be a point of local minimum of the functional J if $J[y] \geq J[z]$ for all y that satisfy the condition $y(x) \equiv z(x)$ for $x \in \Omega/\omega$ and are solutions to the problems (10.4) with δ small enough and some k.

THEOREM 10.1. *Let the functional $J[y]$ with the constraint $F(\dot{y}, y, x) = 0$ attain a local minimum at a regular point $z(x)$ in ω. For $z \in C^2(\overline{\omega}, \mathbb{R}^n)$ and $H_{p^i_j}(\dot{z}(x), z(x), x) \in C^1(\overline{\omega})$ the following assertions hold:*

(a) *there exists a function $f \in C^1(\overline{\omega})$ such that the pair z, f satisfies* (10.6),

(b) *$J''[z](u,u) \geq 0$ for all $u \in J^{k,\varepsilon}(\Gamma, *, *)$,*

(c) *for every point $x \in \omega$ and all vectors $\xi \in \mathbb{R}^n$, $\lambda \in \mathbb{R}^m$, satisfying the orthogonality condition $(A(x)\lambda, \xi)_n = 0$, $A_{ij}(x) = F_{p^i_j}(\dot{z}(x), z(x), x)$, the following inequality holds at x:*

$$\left[H_{p^i_\alpha p^j_\beta}(\dot{z}(x), z(x), x) - f(x)F_{p^i_\alpha p^j_\beta}(\dot{z}(x), z(x), x)\right]\xi_i\xi_j\lambda_\alpha\lambda_\beta \geq 0.$$

PROOF. We define the family of functions $y(x,t)$, $|t| \leq 1$, by the equalities $y(x,t) \equiv z(x)$ for $x \in \Omega/\omega$ and $y(x,t) = z(x) + tu(x) + R[tu](x)$ for $x \in \omega$. A function $\Phi(t) \equiv J[y(x,t)]$ is twice continuously differentiable and attains a local minimum at $t = 0$. Consequently, $\Phi'(0) = 0$ and $\Phi''(0) \geq 0$ for all $u \in J^{k,\varepsilon}(\Gamma, *, *)$. From the first relation it follows that $J'[z]u \equiv 0$, i.e., z is a critical regular point in ω for the functional J with constraint. Therefore, the assertion (a) is obtained from Lemma 10.2. The inequality $\Phi''(0) \geq 0$ coincides with the inequality $J''[z](u,u) \geq 0$, which proves (b). Assertion (c)follows from (b) and Theorem 7.2 from Chapter 1. □

10.6. The method of Lagrange multipliers. This method shows how to obtain a necessary condition for an extremum of a functional with constraint by variation of the auxiliary functional without constraints

$$I[y] = \int\limits_\Omega [H[y] - f(x)F[y]]\,dx + \int\limits_{\partial\Omega} h[y]\,dS. \tag{10.13}$$

The function f in (10.13) is called a Lagrange multiplier.

To justify the method, we consider $\mu \in C^1(\overline{\omega}, \mathbb{R}^n)$ such that $\mu\big|_\Gamma = 0$. The function $\Phi(t) = I[z + t\mu]$ is twice continuously differentiable and

$$\begin{aligned} \Phi'[0] &= \int\limits_\omega [H'[z]\mu - fL\mu]\,dx + \int\limits_S h'[z]\mu\,dS, \\ \Phi''[0] &= \int\limits_\omega [H''[z](\mu,\mu) - fF''[z](\mu,\mu)]\,dx + \int\limits_S h''[z](\mu,\mu)\,dS. \end{aligned} \tag{10.14}$$

Since $\Phi'(0) = 0$, we obtain (10.6). Thus, the method of Lagrange multipliers can be applied to the Lagrange equation and the natural boundary conditions for the functional (10.1) with the constraint (10.3).

The condition $\Phi''(0) \geq 0$ differs from the condition (a) in Theorem 10.1. Namely, $\mu(x)$ does not necessarily satisfy the equation $L\mu = 0$ in ω. The example below shows that the condition $\Phi''[0] \geq 0$ for all $\mu(x)$ is not necessary for a local minimum of a functional with constraint at a regular point $z(x)$ in ω. Let $\overline{\omega} \subset \Omega$ and $H(\dot{y}, y, x) = Q(F(\dot{y}, y, x))$, where $Q(\tau)$ is a smooth function such that $Q(0) = 0$, $Q'(0) = 0$, and $Q''(0) = -1$. It is obvious that $H(\dot{y}, y, x) = 0$ for all y satisfying the constraint. Therefore, any regular point $z(x)$ in ω is a point of local minimum for the functional with constraint; moreover, a function $f \in C^1(\overline{\omega})$ satisfies (10.6). In our case, (10.6) takes the form

$$Q'(0)\left[-\frac{d}{dx_j}F_{p^i_j}(\dot{z}, z, x) + F_{q^i}(\dot{z}, z, x)\right] = L^* f.$$

Since $Q'(0) = 0$, we have $L^* f = 0$. From the definition of a regular point, in the case $S \neq \varnothing$ we obtain $f = 0$. A simple computation yields the equality

$$\Phi''(0) = -2^{-1} \int_\omega (L\mu)^2 \, dx.$$

Hence $\Phi''(0)$ is negative for $L\mu \not\equiv 0$. Thus, the method of Lagrange multipliers gives a wrong answer to the question on necessary conditions based on the nonnegativity of the second variation.

For the functional (10.13) the Legendre–Hadamard condition differs from the condition (c) in Theorem 10.1 by the orthogonality requirement $(A(x)\lambda, \xi)_n = 0$. Let us show that the Legendre–Hadamard condition for the functional (10.13) is not necessary for the functional with constraint to attain a local minimum at a regular point $z(x)$ in ω. We use the example just constructed. In this case (cf. Theorem 7.2(b), Chapter 1) the Legendre–Hadamard condition takes the form $-(A(x)\lambda, \xi)^2 \geq 0$. But the last inequality fails for all $\lambda \in \mathbb{R}^m$ and $\xi \in \mathbb{R}^n$. Thus, the method of Lagrange multipliers leads to a false conclusion if we use it for deriving a necessary condition similar to the Legendre–Hadamard condition for a minimum of a functional with constraint.

10.7. A sufficient condition for a local minimum of the functional with constraint at a critical regular point. Let $z(x)$ be a critical regular point in ω for the functional with constraint and let $z \in C^2(\overline{\omega}, \mathbb{R}^n)$, $H_{p^i_j}(\dot{z}(x), z(x), x) \in C^1(\overline{\omega})$. Then $z(x)$, $f(x)$ satisfy (10.6) and $J''[z](u, u)$ satisfies (10.10).

THEOREM 10.2. *If for all* $u \in J^{k,\varepsilon}(\Gamma, *, *)$

$$\int_\omega [H''[z](u, u) - fF''[z](u, u)] \, dx + \int_S h''[z](u, u) \, dS \geq C\langle u\rangle^2, \quad C > 0, \tag{10.15}$$

where $\langle\cdot\rangle$ *is either* $\|\cdot\|_2$ *or* $\|\cdot\|_{1,2}$, *then* $z(x)$ *is a local minimum point in* $C^{k,\varepsilon}(\overline{\omega}, \mathbb{R}^n)$, *where* $k \geq k_0$ *and* k_0 *is defined in Theorem 7.2 and depends on the choice of the norm* $\langle\cdot\rangle$.

PROOF. By Lemma 10.1 and formula (10.15), we have

$$J[y] - J[z] \geq C\langle u\rangle^2 + O(|u|^3_{1,\varepsilon}) \tag{10.16}$$

for all y satisfying (10.4) and the condition $y(x) \equiv z(x)$ in Ω/ω. Arguing as in Theorem 8.3, from (10.16) we derive the estimate

$$J[y] - J[z] \geq |y-z|^3_{1,\varepsilon}(\omega), \quad |y-z|_{k,\varepsilon}(\omega) < \delta.$$

□

10.8. Variational problems in the mechanics of an incompressible medium. Variational problems in the mechanics of an elastic incompressible medium include the problems of minimizing the energy functional (10.1) defined on the set of functions y satisfying (10.2) and the constraint

$$F(\dot{y}) \equiv \det \dot{y} - 1 = 0. \tag{10.17}$$

The set of all solutions to the problem (10.4) with F taken from (10.17) has surface structure (cf. Theorem 9.3). Hence Theorems 10.1 and 10.2 are true for the constraint (10.17). However, these theorems have an essential drawback. Namely, it is required that $z(x)$ in ω be too smooth. For the constraint (10.17) it is possible to obtain the necessary conditions for $z(x)$ to be an extremal point in the energy space.

Let the functional (10.1) be well defined on functions y belonging to the Sobolev space $W^1_r(\omega, \mathbb{R}^m)$ and satisfying (1.2). We assume that the functional (10.1) is differentiable in the following sense:

(10.18)
$$\int_\Omega [H(\dot{y}+\dot{h}, y, x+\xi) - H(\dot{y}, y, x))]\,dx = \int_\Omega [H_{p^i_j}(\dot{y}, y, x)h^i_{x_j} H_{x_i}(\dot{y}, y, x)\xi^i]\,dx + \beta,$$

$$h \in W^1_r(\omega, \mathbb{R}^m), \quad \xi \in \mathbb{R}^m, \quad \beta = o(\|h\|_{1,r} + |\xi|).$$

THEOREM 10.3. *Let the functional* (10.1) *be defined on functions satisfying the condition* (10.2) *and the constraint* (10.17). *If it attains a local minimum at $z \in W^1_r(\omega, \mathbb{R}^m)$, then for any smooth solenoidal vector-valued function u with compact support the following equality holds*:

$$\int_\Omega \left[H_{p^i_\alpha}(\dot{z}, z, x) z^i_{x_j} u^j_{x_\alpha} - H_{x_j}(\dot{z}, z, x)u^j\right] dx = 0. \tag{10.19}$$

PROOF. We fix a bounded subdomain $\omega \subset \Omega$ with boundary $\partial\omega \in C^{2,\varepsilon}$ and take $\eta \in C^{1,\varepsilon}(\overline{\omega}, \mathbb{R}^m)$ such that

$$\det \dot{\eta}(x) - 1 = 0, \quad x \in \overline{\omega}, \quad \eta(x) - x = 0, \quad x \in \partial\omega, \quad |\eta(x) - x|_{1,\varepsilon} < \delta,$$

where δ is sufficiently small. Then $\eta(x)$ is a diffeomorphism of ω onto itself. The inverse mapping $x = x(\eta)$ satisfies similar conditions:

$$\det \dot{x}(\eta) - 1 = 0, \quad \eta \in \overline{\omega}, \quad x(\eta) - \eta = 0, \quad x \in \partial\omega, \quad |x(h) - \eta|_{1,\varepsilon} < \rho, \tag{10.20}$$

where ρ is sufficiently small. Any solution x to the problem (10.20) can be written as follows (cf. Theorem 9.3):

$$x(\eta) = \eta + u(\eta) + R[u](\eta), \quad R[0] = 0, \quad R'[0] = 0,$$
$$\operatorname{div} u = 0, \quad u|_{\partial\omega} = 0, \quad |u|_{1,\varepsilon} < \delta.$$

We introduce the family $x(\eta, t)$ of diffeomorphisms by the formula

$$x(\eta, t) = \eta + tu(\eta) + R[tu](\eta), \quad t \in [-1, 1]. \tag{10.21}$$

The family of the inverse diffeomorphisms is denoted by $\eta(x, t)$. Consider the family of vector-valued functions $y(x, t) = \{z(x) : x \in \Omega/\omega, z(\eta(x, t)), x \in \omega\}$. Each function of this family belongs to the Sobolev space $W^1_r(\Omega, \mathbb{R}^m)$, satisfies the boundary condition (10.2) and the constraint (10.17), and for $t = 0$ passes through a point of local minimum. We consider the function

$$\Phi(t) = J[y(x, t)] - J[z] = \int\limits_\omega H(\dot{z}(\eta)\dot{\eta}(x, t), z(\eta), x)\, dx - \int\limits_\omega H(\dot{z}(x), z(x), x)\, dx.$$

In the first integral on the right-hand side of the last equality, we make the change of variables $x \to \eta = \eta(x, t)$ in ω, and in the second integral we denote by η the integration variable. Then

$$\Phi(t) = \int\limits_\omega H(\dot{z}(\eta)\dot{x}^{-1}(\eta, t), z(\eta), x(\eta))|\det \dot{x}(\eta, t)|\, d\eta - \int\limits_\omega H(\dot{z}(\eta), z(\eta), \eta))\, d\eta.$$

From (10.21) it follows that

$$x(\eta, t) = \eta + tu + O(t^2), \quad \det \dot{x}(\eta, t) \equiv 1, \quad \dot{x}^{-1}(\eta, t) = I - t\dot{u}(\eta) + O(t^2).$$

By (10.18), $\Phi(t)$ is differentiable with respect to t for $t = 0$ and

$$\Phi'(0) = \int\limits_\omega \left[-H_{p^i_\alpha}(\dot{z}(\eta), z(\eta), \eta) z^i_{\eta_j}(\eta) u^j_{\eta_\alpha}(\eta) + H_{x_i}(\dot{z}(\eta), z(\eta), \eta) u^i\right] d\eta.$$

Since $\Phi(t)$ attains a local minimum at $t = 0$ and, consequently, $\Phi'(0) = 0$, we see that the last equality restricted to the set of all smooth solenoidal compactly supported functions coincides up to notation with (10.19). □

REMARK 10.1. If $z \in C^2(\overline{\omega}, \mathbb{R}^m)$ and $H_{p^i_j}(\dot{z}, z, x) \in C^1(\overline{\omega})$, then (10.19) is equivalent to (10.6), (10.17). Indeed, integrating (10.19) by parts, we obtain

$$\int\limits_\omega \left[\frac{d}{dx_\alpha}(H_{p^i_\alpha}(\dot{z}, z, x) z^i_{x_j}) + H_{x_j}(\dot{z}, z, x)\right] u^j(x)\, dx = 0$$

for all smooth solenoidal vector-valued functions u whose compact supports are contained in ω. Therefore, there exists a function $q \in C^1(\overline{\omega})$ such that

$$\frac{d}{dx_\alpha}\left[H_{p^i_\alpha}(\dot{z}, z, x) z^i_{x_j}\right] + H_{x_j}(\dot{z}, z, x) = q_{x_j}, \quad j = 1, \ldots, m.$$

We rewrite the last equality in the form

$$H_{p^i_\alpha}(\dot z, z, x) z^i_{x_j x_\alpha} + H_{q^i}(\dot z, z, x) z^i_{x_j} + H_{x_j}(\dot z, z, x) \\ + z^i_{x_j}\left[\frac{d}{dx_\alpha} H_{p^i_\alpha}(\dot z, z, x) - H_{q^i}(\dot z, z, x)\right] = q_{x_j}, \quad j = 1, \dots, m.$$

Since the first line is the ordinary differential of $H(\dot z, z, x)$, the last equality is equivalent to the relation

$$z^i_{x_j}\left(-\frac{d}{dx_\alpha} H_{p^i_\alpha} - H_{q^i}\right) = -f_{x_j}, \quad f = q - H, \quad H = H(\dot z, z, x), \quad j = 1, \dots, m.$$

Since $\det \dot z(x) \equiv 1$ in ω, the matrix $\dot z(x)$ is invertible. This yields (10.6) in view of Lemma 9.4. Reversing the above arguments, we reduce (10.6) to (10.19).

Theorem 10.2 provides sufficient conditions for a minimum of the functional with constraint over small smooth perturbations. For the constraint (10.17) there is an example of a critical point at which the functional (10.1) defined on functions satisfying (10.2) attains a local minimum over small smooth perturbations but not over all small perturbations from the energy space.

Introduce the notation

$$\begin{gathered} \omega = \Omega \subset \mathbb{R}^m, \quad \varnothing \neq \Sigma = \Gamma \neq \partial\Omega, \\ H = H(\dot y) \equiv \Phi(\operatorname{tr} g_y), \quad \Phi \in C^3(\mathbb{R}), \quad \Phi'(m) > 0, \\ h(y, x) = \lambda b(y, x), \quad \lambda \in \mathbb{R}, \quad b \in C^3(\mathbb{R}^m \times \mathbb{R}^m), \\ b_{q^i}(x, x) = \nu^i(x), \quad x \in \partial\omega. \end{gathered}$$

The variational problem for such densities describes the equilibrium state of an elastic incompressible rubber-type material under an action that depends on the parameter λ and is applied to the boundary in the normal direction.

For $z(x) \equiv x$ the relations (10.6) take the form

$$\begin{gathered} \nabla f(x) \equiv 0, \quad x \in \omega, \\ \lambda b_{q^i}(x, x) + [2\Phi'(m) - f(x)]\nu^i \equiv 0, \quad i = 1, \dots, m, x \in S. \end{gathered}$$

Consequently, $z(x) \equiv x$ is a critical regular point for the functional of the deformation energy of an incompressible medium with the function $f = \lambda + 2\Phi'(m)$. A solution z of (10.6) such that

$$z(x) \equiv x, \quad f(x) \equiv \lambda + 2\Phi'(m) \tag{10.22}$$

is called a nondeformation state of the elastic incompressible medium. It can be characterized as an exterior action on the part of the boundary S that leaves points of the medium in their places, but creates some pressure $f(x)$ in the medium.

THEOREM 10.4. 1. *There exists λ_0 such that for all $\lambda \in (-\lambda_0, \lambda_0)$ the nondeformation state of an elastic incompressible medium is a point of local minimum of the deformation energy functional over small smooth perturbations.*

2. *Let $m = 2$. There exists a monotone increasing bounded function $\Phi(t)$ such that for some $\lambda_1 \in (-\lambda_0, \lambda_0)$ the nondeformation state of an elastic incompressible medium is not a point of local minimum of the deformation energy over all small perturbations from the energy space.*

PROOF. 1. Using (10.10), for $u \in C^{k,\varepsilon}(\overline{\omega}, \mathbb{R}^m)$ we find that

$$
\begin{aligned}
J''[x](u,u) &= \int\limits_{\omega} \left[2\Phi'(m) u^i_{x_j} u^i_{x_j} + (\lambda + 2\Phi'(m)) u^i_{x_j} u^j_{x_i}\right] dx \\
&+\lambda \int\limits_{\partial\omega} b_{q^i q^j}(x,x) u^i u^j \, dS, \quad \operatorname{div} u = 0, \quad u|_\Gamma = 0.
\end{aligned} \tag{10.23}
$$

Since

$$u^i_{x_j} u^i_{x_j} + u^i_{x_j} u^j_{x_i} = \frac{1}{2}(u^i_{x_j} u^i_{x_j} + 2u^i_{x_j} u^j_{x_i} + u^j_{x_i} u^j_{x_i}) = \frac{1}{2}\sum_{i,j}(u^i_{x_j} + u^j_{x_i})^2,$$

$$|u^i_{x_j} u^j_{x_i}| \le \frac{1}{2}(u^i_{x_j} u^i_{x_j} + u^j_{x_i} u^j_{x_i}) = u^i_{x_j} u^i_{x_j},$$

the expression in brackets on the right-hand side of the first equality in (10.23) is estimated from below in terms of $\Phi'(m)\sum_{i,j}(u^i_{x_j} + u^j_{x_i})^2 - |\lambda| u^i_{x_j} u^i_{x_j}$. Taking into account the Korn inequality

$$\int\limits_{\omega} \sum_{i,j} \left(u^i_{x_j} + u^j_{x_i}\right)^2 dx \ge L \int\limits_{\omega} u^i_{x_j} u^i_{x_j} \, dx, \quad u \in W^1_2(\omega, \mathbb{R}^m), \quad u|_\Gamma = 0,$$

and the boundedness of the embedding of $W^1_2(\omega)$ into $L_2(\partial\omega)$, i.e.,

$$M \int \omega u^i_{x_j} u^i_{x_j} \, dx \ge \int_{\partial\omega} u^2 \, dS,$$

we obtain the estimate

$$J''[x](u,u) \ge [\Phi'(m)L - |\lambda|(1 + \mu M)] \int\limits_{\omega} u^i_{x_j} u^i_{x_j} \, dx, \tag{10.24}$$

where μ is the smallest constant satisfying the inequality

$$|b_{q^i q^j}(x,x)\xi^i \xi^j| \le \mu |\xi|^2, \quad \xi \in \mathbb{R}^m.$$

The required assertion follows from (10.24) for $|\lambda| < \lambda_0 = \Phi'(m)L(1 + \mu M)^{-1}$ and Theorem 10.2.

2. We begin with the construction a special family of mappings. Let S_1 be a circle of unit radius and let $\varphi(\theta)$, $\theta \in S_1$, be a nonnegative smooth scalar function. We define the mapping $y : \mathbb{R}^2 \to \mathbb{R}^2$ by the equality

$$y(x) = \frac{x}{|x|}\left(|x|^2 + \varphi^2\left(\frac{x}{|x|}\right)\right)^{1/2}. \tag{10.25}$$

The mapping y assigns a single point $y(x) \in \mathbb{R}^2$ to a point $x \neq 0$ and takes 0 to the surface $\theta\varphi(\theta)$, $\theta \in S_1$. Let $0 \in \omega$ and $1 \le r < 2$. Then $y \in W^1_r(\omega, \mathbb{R}^2)$ and

$$\|y(x) - x\|^r_{1,r} \le C\big[\|\varphi\|^r_{1,r}(S_1) + \|\varphi\|^{r/2}_r(S_1)\big]. \tag{10.26}$$

From (10.25) it follows that $|y|^2 = |x|^2 + \varphi^2(x|x|^{-1})$. Therefore, $|y|d|y| = |x|d|x|$. Since $y(x)|y(x)|^{-1} = x|x|^{-1} = \theta$, y is volume preserving, i.e., $\det \dot{y}(x) \equiv 1$ for $x \neq 0$. Points on the circle can be parametrized by the polar angle α. Hence $\varphi = \varphi(\alpha)$ is a scalar function of a scalar argument α. We assume that a sector of angle $|\alpha| \le \alpha_0$

cuts off a piece $\sigma \subset S$ of the boundary $\partial\omega$. The piece can be represented as the graph of the function $R = R(\theta)$. It easy to compute that

$$\operatorname{tr} g_y = 2 + (\varphi^2 + \dot\varphi^2)(|x|^{-2} - (|x|^2 + \varphi^2)^{-1}).$$

We fix a piecewise linear function $\varphi_0(\alpha)$ such that

$$\varphi_0(\alpha) \equiv 0 \text{ for } \alpha \notin [-\alpha_0, \alpha_0], \quad |\dot\varphi_0(\alpha)| \equiv 1, \quad 0 \le \varphi_0(\alpha) \le 1$$

and introduce the family of mappings $y_\varepsilon(x)$ according to (10.25) with $\varphi = \varepsilon\varphi_0$. It is obvious that $y_\varepsilon \in W_r^1(\omega, \mathbb{R}^2)$ satisfies the boundary condition (10.2) and the incompressibility condition (10.17). By (10.26), $\|y_\varepsilon(x) - x\|_{1,r} \to 0$ as $\varepsilon \to 0$.

We consider the difference

(10.27)

$$J[y_\varepsilon] - J[x] = \int_{-\alpha_0}^{\alpha_0} d\alpha \int_0^{R(\alpha)} [\Phi(2 + \varepsilon^2(\varphi_0^2(\alpha) + 1)(r^{-2} + \varepsilon^2\varphi_0^2(\alpha))^{-1})) - \Phi(2)]r\, dr$$
$$+ \lambda \int_\sigma [b(\theta(R^2(\alpha) + \varepsilon^2\varphi_0^2(\alpha))^{1/2}, x) - b(x,x)]\, dS$$

and assume that the function $\Phi(t)$ is monotone. Then

$$\Phi\big(2 + \varepsilon^2(\varphi_0^2(\alpha) + 1)\varepsilon^2\varphi_0^2(\alpha) r^{-2}(r^2 + \varepsilon^2\varphi_0^2(\alpha))^{-1}\big) \ge \Phi(2 + 4\varepsilon^4 r^{-4}).$$

Hence the first integral in (10.27) is estimated from above by

$$2\alpha_0 \int_0^R [\Phi(2 + 4\varepsilon^4 r^{-4}) - \Phi(2)]r\, dr,$$

where $R = \operatorname{diam}\omega$. Making the change of variables $4\varepsilon^4 r^{-4} = \xi$, we estimate the integral from above by

$$\alpha_0\varepsilon^2 \int_0^\infty \Phi(2 + \xi) - \Phi(2)]\xi^{-3/2}\, d\xi.$$

Since b is continuously differentiable, we use the last equality in (10.20) to write the second integral in (10.27) as follows:

$$2^{-1}\lambda\varepsilon^2 \int_\sigma \varphi_0^2(\alpha) R(\alpha)^{-1}(\nu, \theta)_2\, dS + \lambda O(\varepsilon^4)$$

as $\varepsilon \to 0$. Combining the above estimates, we arrive at the inequality

$$J[y_\varepsilon] - J[x] \le \varepsilon^2 \bigg[\alpha_0 \int_0^\infty [\Phi(2 + \xi) - \Phi(2)]\xi^{-3/2}\, d\xi$$
$$+ 2^{-1}\lambda \int_\sigma \varphi_0^2(\alpha) R(\alpha)^{-1}(\nu, \theta)_2\, dS\bigg] + \lambda O(\varepsilon^4)$$

as $\varepsilon \to 0$. Keeping $\Phi'(2)$ unchanged, we choose a monotonically increasing bounded density Φ and place the point 0 in the domain ω so that the expression in brackets becomes negative for some $\lambda_1 \in (-\lambda_0, \lambda_0)$. Then for $\lambda = \lambda_1$ and all sufficiently small $\varepsilon \ne 0$ we obtain the inequality $J[y_\varepsilon] < J[x]$, which proves the second part of the theorem. $\square$

Theorem 10.4 describes the case in which a collapse of an elastic medium with the appearence of a cavity is profitable from the energetic point of view, whereas small smooth deformations are impossible.

Appendix

1. Function spaces in a domain ω. Let ω be a bounded domain in $\mathbb{R}^m$, $\overline{\omega}$ its closure, and $\partial\omega$ its boundary. We denote by $C^k(\overline{\omega})$ the Banach space of functions that are k-times continuously differentiable in ω and are continuous in $\overline{\omega}$ together with all their derivatives of order up to k. The norm in $C^k(\overline{\omega})$ is defined as follows:

$$|u|_k = \sup_{x\in\omega} \sum_{|\alpha|\le k} |D^\alpha u(x)|.$$

We use the notation $C^0(\overline{\omega}) \equiv C(\overline{\omega})$ for the space of continuous functions in $\overline{\omega}$.

For $\varepsilon \in [0,1]$ the set $C^{k,\varepsilon}(\overline{\omega})$ of functions $u \in C^k(\overline{\omega})$ such that

$$|u|_{k,\varepsilon} = |u|_k + \sup_{\substack{x,y\in\omega \\ x\ne y}} \sum_{|\alpha|=k} \frac{|D^\alpha u(x) - D^\alpha u(y)|}{|x-y|^\varepsilon} < \infty$$

is a Banach space with the norm $|\cdot|_{k,\varepsilon}$, called the *Hölder space.* The set of functions that are continuous in ω together with all their derivatives of orders up to k is equipped with the norm $|\cdot|_{k,\varepsilon}$ and is denoted by $C^{k,\varepsilon}(\omega)$. It is obvious that $C^{k,0} = C^k$.

The Banach space of q-integrable functions is denoted by $L_q(\omega)$, $q \in [1,\infty)$. The norm in $L_q(\omega)$ is defined as follows:

$$\|u\|_q = \left(\int_\omega |u(x)|^q\,dx\right)^{1/q}.$$

Let a function $u(x)$ belong to $L_q(\omega)$ and have the Sobolev derivatives $D^\alpha u \in L_q(\omega)$, $|\alpha| \le l$, where l is a natural number. The set of all such functions with the norm

$$\|u\|_{l,q} = \left[\sum_{|\alpha|\le l} \int_\omega |D^\alpha u|^q\,dx\right]^{1/q}$$

is a complete space, called the *Sobolev space*, denoted by $W^l_q(\omega)$. For $l=0$ we set $W^0_q(\omega) = L_q(\omega)$.

For $l > 0$ we write $[l]$ for its integer part and introduce the notation $\{l\} = l - [l]$. By $W^l_q(\omega)$ we mean the set of functions $u \in W^{[l]}_q(\omega)$ such that

$$\|u\|_{l,q} = \|u\|_{[l],q} + \sum_{|\alpha|=[l]} \left[\int_\omega\int_\omega \frac{|D^\alpha u(x) - D^\alpha u(y)|^q}{|x-y|^{m+\{l\}q}}\,dx\right]^{1/q} < \infty.$$

This space is called the Slobodetskiĭ space. It is complete in the norm $\|\cdot\|_{l,q}$.

Let $C_0^\infty(\omega)$ denote the space of smooth functions with compact support in ω. Its completion in the $W_q^l(\omega)$-norm is denoted by $\overset{0}{W}{}_q^l(\omega)$.

All the above spaces consist of scalar functions. We say that an n-dimensional vector-valued function belongs to $C^{k,\varepsilon}(\overline{\omega}, \mathbb{R}^n)$ or $W_q^l(\omega, \mathbb{R}^n)$ if each of its components belongs to $C^{k,\varepsilon}(\overline{\omega})$ or $W_q^l(\omega)$ respectively. For the norm of a vector-valued function we can take the sum of the norms of its components or some other expression that provides an equivalent norm. In the space $L_2(\omega, \mathbb{R}^n)$ the inner product can be introduced by the equality

$$\langle u, v \rangle = \int\limits_{\omega} (u(x), v(x))_n \, dx;$$

thus, $L_2(\omega, \mathbb{R}^n)$ becomes a Hilbert space.

2. The boundary of a domain ω. Let $B_\rho(x_0)$ denote the open ball in $\mathbb{R}^m$ of center x_0 and radius ρ. A one-to-one mapping ψ of $B_\rho(x_0)$ to $\psi(B_\rho(x_0)) \equiv D \subset \mathbb{R}^m$ is called a $C^{k,\varepsilon}$-diffeomorphism of the ball $B_\rho(x_0)$ if ψ is of class $C^{k,\varepsilon}(\overline{B}_\rho(x_0), \mathbb{R}^m)$ and the Jacobi matrix of ψ is nonsingular. By the implicit function theorem, we have $\psi^{-1} \in C^{k,\varepsilon}(\overline{D}, \mathbb{R}^m)$.

DEFINITION 1. We say that the boundary $\partial\omega$ of a domain ω is of class $C^{k,\varepsilon}$ and write $\partial\omega \in C^{k,\varepsilon}$, where k is a natural number and $\varepsilon \in [0, 1]$, if for every point $x_0 \in \partial\omega$ there exist $\rho = \rho(x_0)$ and a diffeomorphism $\psi \in C^{k,\varepsilon}(\overline{B}_\rho(x_0), \mathbb{R}^m)$ such that $\psi(B_\rho(x_0) \cap \omega) \subset \mathbb{R}_+^m \equiv \{x \in \mathbb{R}^m : x_m > 0\}$ and $\psi(B_\rho(x_0) \cap \partial\omega) \subset \partial\mathbb{R}_+^m$.

Thus, the boundary $\partial\omega$ of ω is of class $C^{k,\varepsilon}$ if for every point on $\partial\omega$ there is a neighborhood in which the boundary $\partial\omega$ can be represented as the graph of a $C^{k,\varepsilon}$-function of $m-1$ variables taken from the full collection of variables $x_1, \ldots, x_m$. Indeed, if it is the graph of the function $x_i = f(x_1, \ldots, x_{i-1}, x_{i+1}, \ldots, x_m)$, then $y = \psi(x)$ can be defined as follows:

$$y_j = x_j, \quad j \neq i, \qquad y_i = x_i - f(x_1, \ldots, x_{i-1}, x_{i+1}, \ldots, x_m).$$

The converse assertion is also true. If $\partial\omega \in C^{k,\varepsilon}$, then for every point on $\partial\omega$ there is a neighborhood in which $\partial\omega$ is the graph of a $C^{k,\varepsilon}$-function of $m-1$ variables. Indeed, let $x_0 \in \partial\omega$ and let $\psi(x)$ be a $C^{k,\varepsilon}$-diffeomorphism of $B_\rho(x_0)$. Since the Jacobi matrix of ψ is nonsingular at x_0, the derivatives $\psi_{mx_i}(x_0)$, $i = 1, \ldots, m$, do not vanish simultaneously. Assume that $\psi_{mx_i}(x_0) \neq 0$ for some i. Using the implicit function theorem and the equality $\psi_m(x) = 0$ determining $\partial\omega \cap B_\rho(x_0)$, in a neighborhood of x_0 we can express the variable x_i in terms of the remaining variables as a function f of class $C^{k,\varepsilon}$:

$$x_i = f(x_1, \ldots, x_{i-1}, x_{i+1}, \ldots, x_m).$$

If $\partial\omega \in C^{k,\varepsilon}$ and $k \geq 2$, then the distance function $\rho(x) = \operatorname{dist}(x, \partial\omega)$ is uniquely determined in a sufficiently small neighborhood of $\partial\omega$. We assign to this function a sign by setting $d(x) = \rho(x)$ for $x \notin \overline{\omega}$ and $d(x) = -\rho(x)$ for $x \in \omega$. Then the function $d(x)$ is of class $C^{k,\varepsilon}$ in a sufficiently small neighborhood of $\partial\omega$ (cf. [**10**]). The function $d(x)$ is referred to as the *oriented distance.* Sometimes it is convenient to work with a function of class $C^{k,\varepsilon}(\mathbb{R}^m)$ that coincides with $d(x)$ in a small neighborhood of $\partial\omega$.

Let $\overline{\omega} \subset B_R(0)$ and $\partial\omega \in C^{k,\varepsilon}$. We recall that the extension operator Π acting from a space of functions in ω to a space of functions in $B_R(0)$ is characterized by the equality $(\Pi u)(x) = u(x)$ for $x \in \omega$. From the Whitney construction [**36**] we obtain the following assertion.

THEOREM 1 (on extension of functions defined in ω). *There exists a linear extension operator* $\Pi : C^{k,\varepsilon}(\overline{\omega}) \to C^{k,\varepsilon}(\overline{B}_R(0))$ *satisfying the estimates*

$$|\Pi u|_{s,\varepsilon} \leq C_s |u|_{s,\varepsilon}, \quad \|\Pi u\|_{l,q} \leq C_{lq} \|u\|_{l,q}$$

for all $s = 1, \dots, k$, $q \in [1, \infty)$, *and* $l \in [0, k]$; *moreover, all the functions* $(\Pi u)(x)$ *vanish outside some neighborhood of* ω.

If $\partial\omega \in C^{k,\varepsilon}$, then the set of restrictions to ω of infinitely differentiable functions defined in $B_\rho(0)$ is dense in $W_q^k(\omega)$ and $C^{k,\varepsilon}(\overline{\omega})$ for $\varepsilon = 0$. However, this set is not dense in $C^{k,\varepsilon}(\overline{\omega})$ for $\varepsilon \neq 0$. The space $C^{k,\mu}(\overline{\omega})$ is not dense in $C^{k,\varepsilon}(\overline{\omega})$ for any $0 < \varepsilon < \mu \leqslant 1$ (cf. [**15**]).

3. Function spaces on $\partial\omega$. Let $x_j \in \partial\omega$, $j = 1, \dots, N$, be a finite collection of points such that the balls $B_\rho(x_j)$ cover the boundary $\partial\omega$. We fix the set of functions $\varphi_j \in C_0^\infty(B_\rho(x_j))$ such that $\sum_{j=1}^N \varphi_j(x) \equiv 1$ on $\partial\omega$. Any function $u(x)$ defined on $\partial\omega$ can be represented in the form $u(x) = \sum_{j=1}^N u_j(x)$, where $u_j(x) = u(x)\varphi_j(x)$. Let $\partial\omega \in C^{k,\varepsilon}$ and let $\psi_j(x)$ be the diffeomorphisms of the balls $B_\rho(x_j)$ from Definition 1.

DEFINITION 2. (a) We say that a function u is of class $C^{s,\varepsilon}(\partial\omega)$, $s = 1, \dots, k$, if $u_j(\psi_j^{-1}(y)) \in C^{s,\varepsilon}(\mathbb{R}^{m-1} \cap \overline{\psi}_j(B_\rho(x_j)))$ for $y \in \mathbb{R}^{m-1} \cap \psi(B_\rho(x_j))$. The $C^{s,\varepsilon}(\partial\omega)$-norm of u is defined as follows:

$$|u|_{s,\varepsilon} = \sum_{j=1}^N |u_j \circ \psi_j^{-1}|_{s,\varepsilon}.$$

(b) We say that a function u is of class $W_q^l(\partial\omega)$, $0 \leq l \leq k$, if $u(\psi_j^{-1}(y)) \in W_q^l(\mathbb{R}^{m-1} \cap \psi_j(B_\rho(x_j)))$ for $y \in \mathbb{R}^{m-1} \cap \psi_j(B_\rho(x_j))$. The $W_q^l(\partial\omega)$-norm of u is defined as follows:

$$\|u\|_{l,q} = \sum_{j=1}^N \|u_j \circ \psi_j^{-1}\|_{l,q}.$$

One can show (cf. [**7, 36, 37**]) that u belongs to the above spaces independently of the choice of points x_j, elements of the partition of unity $\varphi_j(x)$, and diffeomorphisms $\psi_j(x)$. The choice of some other parameters only leads to the replacement of the fixed norms of u by equivalent norms. In particular, we can use the diffeomorphism ψ corresponding to the function f whose graph locally describes the boundary $\partial\omega$. In this case,

$$u_j \circ \psi_j^{-1}(y) = u_j(y_1, \dots, y_{i-1}, f(y_1, \dots, y_{i-1}, y_{i+1}, \dots, y_m), y_{i+1}, \dots, y_m).$$

If $\partial\omega \in C^{k+1,\varepsilon}$, then $\partial\omega$ is locally represented as the graph of a function f of class $C^{k+1,\varepsilon}$, i.e., $x_i = f(x_1, \dots, x_{i-1}, x_{i+1}, \dots, x_m)$. The unit outward normal $\nu(x)$ to

$\partial\omega$ is given up to sign by the expression

$$(1+|\nabla f|^2)^{-1/2}(f_{x_1},\dots,f_{x_{i-1}},1,f_{x_{i+1}},\dots,f_{x_m}),$$
$$f=f(x_1,\dots,x_{i-1},x_{i+1},\dots,x_m).$$

The vector field $\nu\circ\psi^{-1}$ corresponding to the diffeomorphism generated by f is given by the same expression with x_l replaced by y_l, $l\neq i$. Therefore, $\nu\in C^{k,\varepsilon}(\partial\omega,\mathbb{R}^m)$. The equality $\nu(x)=\nabla d(x)$ holds at $x\in\partial\omega$. The inward and outward normal derivatives of the function $\rho(x)$ satisfy the equalities $\nu(x)=-\nabla\rho(x)$ and $\nu(x)=\nabla\rho(x)$ respectively. The second order derivatives of $d(x)$ at $x\in\partial\omega$ satisfy the relation

$$d_{x_ix_j}(x)\xi_i\xi_j=-K_{ij}(x)\xi_i'\xi_j',$$

where $K(x)$ denotes the matrix of principal curvatures of $\partial\omega$ at x and $\xi'=\xi-\nu(\xi,\nu)_m$ is the tangent component of $\xi\in\mathbb{R}^m$.

We say that a function u is of class $C^{k,\varepsilon}_{\text{loc}}(\Omega)$, $\Omega\subset\mathbb{R}^m$, if $u\in C^{k,\varepsilon}(\overline{\omega})$ for any bounded subdomain $\omega\subset\overline{\omega}\subset\Omega$.

We need the following assertion.

THEOREM 2 (on extension of a function from $\partial\omega$). *Let $\partial\omega\in C^{k+1,\varepsilon}$, $k\geq 1$, $\varepsilon\in[0,1]$, and let $d(x)$ be the oriented distance from $x\in\mathbb{R}^m$ to $\partial\omega$. There exists a linear extension operator Π: $C^{k,\varepsilon}(\partial\omega)\to C^{k,\varepsilon}_{\text{loc}}(\mathbb{R}^m)$ such that*

(a) $\Pi u\in C^{k+1,\varepsilon}_{\text{loc}}(\mathbb{R}^m/\partial\omega)$, $u\in C^{k,\varepsilon}(\partial\omega)$;

(b) *the supports of all functions Πu, $u\in C^{k,\varepsilon}(\partial\omega)$, lie in a common sufficiently small neighborhood of $\partial\omega$;*

(c) $d(x)\nabla(\Pi u)(x)\in C^{k,\varepsilon}_{\text{loc}}(\mathbb{R}^m,\mathbb{R}^m)$, $u\in C^{k,\varepsilon}(\partial\omega)$;

(d) *for all $s=1,\dots,k$ and $q\in(1,\infty)$ the following estimates hold:*

$$|\Pi u|_{s,\varepsilon}+|d\nabla\Pi u|_{s,\varepsilon}\leq C_s|u|_{s,\varepsilon},$$
$$\|\Pi u\|_{1,q}+\|d\nabla\Pi u\|_{1,q}\leq C_q\|u\|_{1-1/q,q},$$

where the norms on the left-hand side are taken over the ball $B_K(0)\subset\mathbb{R}^m$ with any K.

PROOF. 1. We begin with the following special case. Let $\omega\subset\mathbb{R}^m$ be the half-ball defined by the relations $|y|\leq 2R$, $y_m\geq 0$. We construct the required operator Π for a function $u(y')$, $y'=(y_1,\dots,y_{m-1})$, that belongs to $C^{k,\varepsilon}(\mathbb{R}^{m-1})$ and has support in the ball of radius R. The extension operator is given by the formula (cf. [**36**])

$$(\Pi u)(y',y_m)=\eta(y_m)\int\limits_{\mathbb{R}^{m-1}}u(y'+z'y_m)K(z')\,dz',$$

where a smooth function $\eta(y_m)$ with compact support is equal to 0 for $|y_m|\geq\delta$ and 1 for $|y_m|\leq\delta/2$, the support of $K\in C_0^\infty(\mathbb{R}^{m-1})$ lies in the ball of center 0 and radius δ; moreover, the integral of $K(z')$ is equal to 1. For sufficiently small δ the support of Πu lies in the ball $B_{2R}(0)$ and is concentrated in the strip of width δ along the hyperplane $y_m=0$. The operator Π is an extension operator, i.e., $(\Pi u)(y',0)=u(y')$. Changing the integration variable, we can prove that the function $(\Pi u)(y',y_m)$ is infinitely differentiable for $y_m\neq 0$. In addition,

$(\Pi u)(y', y_m) \in C^{k,\varepsilon}_{\text{loc}}(\mathbb{R}^m)$. Since $k \geq 1$, the first order derivatives of Πu exist and satisfy the relations

$$y_m \frac{\partial (\Pi u)(y', y_m)}{\partial y_i} = \delta_{im} \frac{\partial \eta(y_m)}{\partial y_i} \int\limits_{\mathbb{R}^m} u(y' + z' y_m) K(z')\, dz' + \eta(y_m) \int\limits_{\mathbb{R}^m} u(y' + z' y_m) K_i(z')\, dz',$$

$$K_i(z') = -\frac{\partial K(z')}{\partial z_i}, \quad i \neq m, \qquad K_m(z') = -\sum_{j=1}^{m-1} \frac{\partial}{\partial z_j}(z_j K(z')).$$

Since $d(y) = -y_m$, the first estimate in the theorem is obvious. The second is contained in [**36**].

Consider a matrix $A(y) = \{a_{ij}(y)\}$ for sufficiently small $|y_m|$ and $a_{ij} \in C^{k,\varepsilon}$. Let a function $h(y', y_m)$ be also defined for $|y_m|$ small enough, belong to $C^{k+1,\varepsilon}$ and satisfy the condition $h(y', 0) = 0$. Then

$$y_m^{-1} h(y', y_m) = \int\limits_0^1 h_{y_m}(y', t y_m)\, dt$$

is of class $C^{k,\varepsilon}$. In the notation $\|\cdot\| = \|\cdot\|_{1,q}$ or $\|\cdot\| = |\cdot|_{s,\varepsilon}$, we obtain the relation

$$\|h(y', y_m) A \nabla \Pi u\| = \|y_m y_m^{-1} h(y', y_m) A \nabla \Pi u\| \leq C \|y_m \nabla \Pi u\|,$$

which allows us to replace $d(y)\nabla(\Pi u)$ by $h(y', y_m) A(y) \nabla(\Pi u)$ in the estimates of the theorem.

2. To study the general case, we consider the balls $B_\rho(x_j)$, the partition of unity $\varphi_j(x)$, and the family of diffeomorphisms $\psi_j(x)$, $j = 1, \dots, N$, from Definition 2. It suffices to construct the extension operator with the required properties only for functions of the form $\varphi_j(x) u(x)$ and define Πu as $\sum_j \Pi(\varphi_j u)$. In the ball $B_\rho(x_j)$, we make the change of variables by means of the diffeomorphism $y = \psi_j(x)$. Since such a change of variables is of class $C^{k+1,\varepsilon}$, we have

$$\frac{\partial}{\partial x_i} = a_{ij}(y) \frac{\partial}{\partial y_j}, \qquad a_{ij} \in C^{k,\varepsilon}(\overline{B_\rho(x_j)}),$$

and, in the new coordinates, $h(y) \equiv d(x(y))$ is of class $C^{k+1,\varepsilon}$ for sufficiently small $|y_m|$. Let R be such that

$$\psi_j(B_\rho(x_j)) \subset B_{2R}(0) \subset \mathbb{R}^m,$$
$$\operatorname{supp}[(\varphi_j u) \circ \psi_j^{-1}] \subset B_R(0) \subset \mathbb{R}^{m-1}, \quad j = 1, \dots, N.$$

We choose ρ small enough. Then the diffeomorphism $\psi_j(x)$ can be regarded as the restriction to $B_\rho(x_j)$ of the diffeomorphism $\widehat{\psi}_j$ which acts from a wider ball $B_{\rho'}(x_j)$ to the domain $\widehat{\psi}_j(B_{\rho'}(x_j)) \supset B_{2R}(0)$ and has the same properties. Let $(\varphi_j u) \circ \psi_j^{-1}$ vanish for $|y'| \geq R$ and let Π_0 denote the extension operator introduced in the first part of the proof. The expression of $\Pi(\varphi_j u)$ is given by the equality

$$\Pi(\varphi_j u) = \{\Pi_0[(\varphi_j u) \circ \psi_j^{-1}]\} \circ \widehat{\psi}_j.$$

We use the estimates which appear after the change of variables and the above estimates to obtain the inequalities

$$\begin{aligned}&|\Pi(\varphi_j u)|_{s,\varepsilon} + |d\nabla\Pi(\varphi_j u)|_{s,\varepsilon}\\ &\quad\le C'\left|\Pi_0[(\varphi_j u)\circ\psi_j^{-1}]\right|_{s,\varepsilon} + C'\left|h(y)A(y)\nabla\Pi[(\varphi_j u)\circ\psi_j^{-1}]\right|_{s,\varepsilon}\\ &\quad\le \left|(\varphi_j u)\circ\psi_j^{-1}\right|_{s,\varepsilon},\end{aligned}$$

where, in the parts of this expression, the norms are taken over ω, the ball $B_{2R}\subset\mathbb{R}^m$, and the hyperplane $y_m = 0$ respectively. By the definition of $C^{s,\varepsilon}(\partial\omega)$, the right-hand side of the last inequality is equal to $|\varphi_j u|_{s,\varepsilon}$. Thus, we have established the first estimate for the function $\varphi_j u$. A similar argument leads to the second estimate. □

The following theorem can be proved with the help of a construction close to that in [**36**].

THEOREM 3 (on extension of a function and its gradient from $\partial\omega$ to $\mathbb{R}^m$). *Let $\partial\omega\in C^2$, $\varphi_0(x)\in C^2(\partial\omega)$, and $\varphi_1(x)\in C^1(\partial\omega)$. There exists a function $\varphi\in C^2(\mathbb{R}^m)$ such that its support lies in a small neighborhood of $\partial\omega$ and the following relations hold:*

$$\varphi(x)=\varphi_0(x),\qquad \frac{\partial\varphi(x)}{\partial\gamma(x)}=\varphi_1(x),\quad x\in\partial\omega,$$

where $\gamma\in C^1(\partial\omega,\mathbb{R}^m)$ is a vector field different from zero and not tangent to $\partial\omega$. If $\operatorname{supp}\varphi_i\subset\partial\omega\cap B_\rho(x_0)$, $i=1,2$, then $\operatorname{supp}\varphi\subset B_{2\rho}(x_0)$.

4. Multiplicative inequalities. We need the following lemma.

LEMMA. *In a bounded domain $\omega\subset\mathbb{R}^m$ with boundary $\partial\omega$ of class C^k, every function $u\in C^{k,\varepsilon}(\overline{\omega},\mathbb{R}^n)$ satisfies the inequalities*

$$|u|_{1,\varepsilon}^{3b}\le C\|u\|_{2b}^{2b}\,|u|_{k,\varepsilon}^{b},\qquad k\ge 3(2+m/2b),\quad b\ge 1/2, \tag{1}$$

$$|u|_{1,\varepsilon}^{3b}\le C\|u\|_{1,2b}^{2b}\,|u|_{k,\varepsilon}^{b},\qquad k\ge 4+3m/2b,\quad b\ge 1/2, \tag{2}$$

where $C=C(k,b)$ is a positive constant.

PROOF. For $u\in C^r(\overline{\omega})$ we use the well-known inequality [**36**]

$$\sum_{|\beta|=\lambda}\sup_{x\in\omega}|D^\beta u|\le C_1\left[\sum_{|\alpha|=r}\left\|D^\beta u\right\|_a\right]^{1-\theta/r}\|u\|_a^{\theta/r}+C_2\|u\|_a, \tag{3}$$

where $\theta\equiv r-\lambda-m/a>0$, $0\le\lambda<r$, $\partial\omega\in C^r$, $a\ge 1$, $C_i>0$, $i=1,2$. Substituting $a=2b$, $b\ge 1/2$, in (3) and raising both sides of the inequality to the power $3b$, we find, after roughening, the relation

$$\sum_{|\beta|=\lambda}\sup_{x\in\omega}|D^\beta u|^{3b}\le C\|u\|_{2b}^{2b}\left[\|u\|_{r,2b}^{3b(1-\theta/r)}\|u\|_{2b}^{3b\theta/r-2b}+\|u\|_{2b}^{b}\right], \tag{4}$$

where C is a positive constant. If the exponent $3b\theta/r-2b$ is nonnegative, then we replace all the norms in brackets by their majorant, i.e., the $C^{r,\varepsilon}$-norm of $u\in$

$C^{r,\varepsilon}(\overline{\omega})$, and obtain the following inequality:

$$\sum_{|\beta|=\lambda} \sup_{x\in\omega} \left|D^\beta u\right|^{3b} \le C\|u\|_{2b}^{2b} \, |u|_{r,\varepsilon}^{b}. \tag{5}$$

The condition that the exponent is nonnegative is equivalent to the inequality $\theta \ge 2r/3$, which implies that θ is positive. Substituting θ from (3), we find the condition on r:

$$r \ge 3(\lambda + m/2b). \tag{6}$$

This condition is stronger than the requirement $r > \lambda \ge 0$. From (5) and (6) for $\lambda = 0, 1, 2$ it follows that

$$\sup_{\omega} |u|^{3b} \le C\|u\|_{2b}^{2b} \, |u|_{r,\varepsilon}^{b}, \quad r \ge 3m/2b, \tag{7}$$

$$\sum_{|\beta|=1} \sup_{\omega} \left|D^\beta u\right|^{3b} \le C\|u\|_{2b}^{2b} \, |u|_{r,\varepsilon}^{b}, \quad r \ge 3(1 + m/2b), \tag{7$'$}$$

$$\sum_{|\beta|=2} \sup_{\omega} \left|D^\beta u\right|^{3b} \le C\|u\|_{2b}^{2b} \, |u|_{r,\varepsilon}^{b}, \quad r \ge 3(2 + m/2b). \tag{7$''$}$$

Combining these inequalities, we arrive at the estimates

$$|u|_0^{3b} \le C\|u\|_{2b}^{2b} \, |u|_{r,\varepsilon}^{b}, \quad r \ge 3m/2b, \tag{8}$$

$$|u|_1^{3b} \le C\|u\|_{2b}^{2b} \, |u|_{r,\varepsilon}^{b}, \quad r \ge 3(1 + m/2b), \tag{8$'$}$$

$$|u|_2^{3b} \le C\|u\|_{2b}^{2b} \, |u|_{r,\varepsilon}^{b}, \quad r \ge 3(2 + m/2b). \tag{8$''$}$$

For $r = k$, (8$''$) implies (1). Substituting u_{x_i} for u in (8$'$), we obtain the inequality

$$\left|u_{x_i}\right|_1^{3b} \le C\|u\|_{1,2b}^{2b} \, |u|_{r+1,\varepsilon}^{b}, \quad r \ge 3(1 + m/2b), \tag{9}$$

for all $u \in C^{r+1,\varepsilon}(\overline{\omega})$. Hence (8) and (9) yield

$$|u|_2^{3b} \le C\|u\|_{1,2b}^{2b} \, |u|_{r+1,\varepsilon}^{b}, \quad r \ge 3(1 + m/2b). \tag{10}$$

For $r + 1 = k$, (10) implies (2). □

5. Function spaces on a part of $\partial\omega \in C^{k,\varepsilon}$. We consider an open set Γ on the boundary $\partial\omega$ and denote by $\overline{\Gamma}$ the closure of Γ. Let the boundary $\partial\Gamma$ of Γ consist of a finite number of connected components.

DEFINITION 3. We say that $\partial\Gamma$ is of class $C^{k,\varepsilon}$ if for every point $x_0 \in \partial\Gamma$ there exist a number $\rho = \rho(x_0)$ and a diffeomorphism $\psi \in C^{k,\varepsilon}(\overline{B_\rho(x_0)}, \mathbb{R}^m)$ with nonsingular Jacobi matrix $\overline{B_\rho(x_0)}$ such that

$$\psi(B_\rho(x_0) \cap \omega) \subset \mathbb{R}_+^m, \quad \psi(B_\rho(x_0) \cap \partial\omega) \subset \partial\mathbb{R}_+^m,$$
$$\psi(B_\rho(x_0) \cap \Gamma) \subset \partial\mathbb{R}_+^{m-1}.$$

DEFINITION 4. Let $B_\rho(x_j)$, $\varphi_j(x)$, and $\psi_j(x)$, $x_j \in \partial\omega$, $j = 1, \dots, N$, be the balls, elements of the partition of unity, and diffeomorphisms from Definition 2. We say that a function u is of class $C^{k,\varepsilon}(\overline{\Gamma})$, $s = 0, \dots, k$, if

$$(u\varphi_j) \circ \psi_j^{-1}(y) \in C^{s,\varepsilon}((\overline{B_\rho(x_j) \cap \Gamma})), \quad y \in \psi_j(B_\rho(x_j) \cap \Gamma).$$

The $C^{s,\varepsilon}(\overline{\Gamma})$-norm of u is defined as follows:

$$|u|_{s,\varepsilon} = \sum \left|(u\varphi_j) \circ \psi_j^{-1}\right|_{s,\varepsilon}.$$

We say that a function u is of class $W_q^l(\Gamma)$, $0 \leq l \leq k$, if

$$(u\varphi_j) \circ \psi_j^{-1}(y) \in W_q^l(\psi_j(B_\rho \cap \Gamma)), \quad y \in \psi_j(B_\rho \cap \Gamma).$$

The $W_q^l(\Gamma)$-norm of u is defined as follows:

$$\|u\|_{l,q} = \sum \left\|(u\varphi_j) \circ \psi_j^{-1}\right\|_{l,q}.$$

The sum in the definitions of norms is taken over those j for which $B_\rho(x_j) \cap \Gamma \neq \varnothing$.

THEOREM 4 (on extension of a function from Γ to $\partial\omega$). *Let $\partial\omega \in C^{k,\varepsilon}$ and let an open set $\Gamma \subset \partial\omega$ with boundary $\partial\Gamma \in C^{k,\varepsilon}$ have a finite number of connected components. There exists a linear extension operator $\Pi : C^{k,\varepsilon}(\overline{\Gamma}) \to C^{k,\varepsilon}(\partial\omega)$ satisfying the estimates $|\Pi u|_{s,\varepsilon} \leq C_s |u|_{s,\varepsilon}$, $\|\Pi u\|_{l,q} \leq C_{lq} \|u\|_{l,q}$ for $s = 1, \dots, k$, $l \in [0, k]$ and $q \in [1, \infty)$; moreover, the supports of all functions Πu are contained in a common sufficiently small neighborhood of Γ.*

PROOF. The arguments are based on the Whitney construction. □

Thus, we have introduced the function spaces $C^{s,\varepsilon}$ and W_q^l on ω, $\partial\omega$, and $\Gamma \subset \partial\omega$. If it is not clear from the context, the norm sign is supplied with the subscript indicating the corresponding space, e.g., $\|\cdot\|_{(\omega)}$, $\|\cdot\|_{(\partial\omega)}$, $\|\cdot\|_{(\Gamma)}$. If $S = \partial\omega / \overline{\Gamma}$, then $\|u\|_{(\partial\omega)} = \|u\|_{(\Gamma)} + \|u\|_{(S)}$.

The proof of the embedding theorems used here can be found in [**17, 36, 37**]. The necessary information about elliptic operators is contained in [**7, 17, 37**].

References

1. I. Ya. Bakel′man, A. A. Verner, and B. E. Kantor, *Introduction to differential geometry "in the large"*, "Nauka", Moscow, 1973. (Russian)
2. M. E. Bogovskiĭ, *Solution of the first boundary-value problem for the equation of continuity of an incompressible medium*, Dokl. Akad. Nauk SSSR **248** (1979), 1037–1040; English transl. in Soviet Math. Dokl. **20** (1979).
3. ———, *A solution of some problems in vector analysis for the operators* div *and* rot, Theory of Cubature Formulas and Applications of Functional Analysis to Problems of Mathematical Physics (Proc. Sem. S. L. Sobolev, Part 1), Inst. Mat. Sibirsk. Otdel. Akad. Nauk SSSR, Novosibirsk, 1980, pp. 5–40. (Russian)
4. J. M. Ball, *Convexity conditions and existence theorems in nonlinear elasticity*, Arch. Rational Mech. Anal. **63** (1977), 337–403.
5. H. Cartan, *Formes différentielles. Calcul différentielle*, Hermann, Paris, 1967; English transl., Houghton-Mifflin, Boston, MA, 1970, 1971.
6. D. G. Ebin and J. Marsden, *Groups of diffeomorphisms and the motion of an incompressible fluid*, Ann. of Math. (2) **92** (1970), 102–163.
7. D. Gilbarg and N. Trudinger, *Elliptic partial differential equations of second order*, Springer-Verlag, Berlin, 1983.
8. G. M. Fikhtengol′ts, *Differential and integral calculus.* I, 7th ed., "Nauka", Moscow, 1969; German transl., VEB Deutscher Verlag Wiss., Berlin, 1972.
9. E. Giusti, *Minimal surfaces and functions of bounded variation*, Birkhäuser, Boston, MA, 1984.
10. A. N Guz′, *Stability of elastic bodies under finite deformations*, "Naukova Dumka", Kiev, 1973. (Russian)
11. ———, *Stability of elastic bodies under comprehensive pressure*, "Naukova Dumka", Kiev, 1979. (Russian)
12. L. Hörmander, *The analysis of linear partial differential operators.* I, II, Springer-Verlag, Berlin, 1983.
13. A. D. Ioffe and V. M. Tikhomirov, *Theory of extremal problems*, "Nauka", Moscow, 1974; English transl, North-Holland, Amsterdam, 1978.
14. L. V. Kapitanskiĭ and K. I. Piletskas, *Some problems of vector analysis*, Zap. Nauchn. Sem. Leningrad Otdel. Mat. Inst. Steklov (LOMI) **138** (1984), 65–85; English transl. in J. Soviet Math. **32** (1986), no. 5.
15. S. G. Kreĭn, Yu. I. Petunin, and E. M. Semenov, *Interpolation of linear operators*, "Nauka", Moscow, 1978; English transl., Amer. Math. Soc., Providence, RI, 1982.
16. O. A. Ladyzhenskaya, *Mathematical questions of the dynamics of a viscous incompressible fluid*, 2nd ed., "Nauka", Moscow, 1970; English transl. of 1st ed., *The mathematical theory of viscous incompressible flow*, Gordon and Breach, New York, 1963.
17. O. A. Ladyzhenskaya and V. A. Solonnikov, *Some problems of vector analysis and generalized statements of boundary-value problems for the Navier–Stokes equations*, Zap. Nauchn. Sem. Leningrad. Otdel. Mat. Inst. Steklov (LOMI) **59** (1976), 81–116; English transl. in J. Soviet Math. **10** (1978), no. 2.
18. O. A. Ladyzhenskaya and N. N. Ural′tseva, *Linear and quasilinear elliptic equations*, 2nd ed., "Nauka", Moscow, 1973; English transl. of 1st ed., Academic Press, New York, 1968.
19. A. I. Lur′e, *Elasticity theory*, "Nauka", Moscow, 1970. (Russian)
20. L. A. Lyusternik and V. I. Sobolev, *A short course in functional analysis*, "Vyssh. Shkola", Moscow, 1982. (Russian)

21. S. G. Mikhlin, *Multidimensional singular integrals and integral equations*, Pergamon Press, Oxford, 1965.
22. C. B. Morrey, *Multiple integrals in the calculus of variations*, Springer-Verlag, Berlin, 1966.
23. L. Nirenberg, *Topics in nonlinear functional analysis*, Courant Inst. Math. Sci., New York Univ., New York, 1974.
24. V. G. Osmolovskiĭ, *An incompressibility condition for a certain class of integral functionals.* I, Zap. Nauchn. Sem. Leningrad. Otdel. Mat. Inst. Steklov (LOMI) **115** (1982), 203–214; English transl. in J. Soviet Math. **28** (1985), no. 5.
25. ______, *An incompressibility condition for a certain class of integral functionals.* II, Problemy Mat. Anal. **9** (1982), 203–214; English transl. in J. Soviet Math. **35** (1986), no. 1.
26. ______, *The Legendre–Hadamard conditions in variational problems for integral functionals on the set of mappings that preserve measure*, Vestnik Leningrad. Univ. **1984**, no. 7 (Ser. Mat. Makh. Astr., vyp. 2), 32–38; English transl. in Vestnik Leningrad. Univ. Math. **17** (1984).
27. ______, *A local description of the set of solutions to a boundary-value problem for a first order nonlinear differential equation*, Vestnik Leningrad. Univ. **1985**, no. 15 (Ser. Mat. Mekh. Astr., vyp. 3), 22–29; English transl. in Vestnik Leningrad. Univ. Math. **18** (1985).
28. ______, *Necessary conditions for the extremum of a functional with constraints in the form of an equality with first derivatives*, Vestnik Leningrad. Univ. **1986**, no. 1 (Ser. Mat. Makh. Astr., vyp. 1), 47–53; English transl. in Vestnik Leningrad. Univ. Math. **19** (1986).
29. ______, *The conjunction problem and the weak form of a necessary condition for the extremum of a functional with constraint*, Vestnik Leningrad. Univ. Mat. Mekh. Astronom. **1986**, no. 1 (Ser. Mat. Mekh. Astr., vyp. 1), 23–29; English transl. in Vestnik Leningrad. Univ. Math. **19** (1986).
30. ______, *The local structure of the set of solutions of a first order nonlinear boundary-value problem with restrictions at the points*, Sibirsk. Mat. Zh. **27** (1986), no. 5, 140–154; English transl. in Siberian Math. J. **27** (1986).
31. ______, *Variational problems on the appearence of a cavity-type discontinuity in a plane problem in the elasticity theory of an incompressible fluid*, Problems in Collapse Mechanics, Kalinin, 1987, pp. 165–172. (Russian)
32. ______, *Rigidity of a surface under deformations satisfying first-order nonlinear differential equations*, Trudy Mat. Inst. Steklov. **179** (1988), 165–172; Engllish transl. in Proc. Steklov Inst. Math. **1989**, no. 2 (179).
33. ______, *A necessary condition for nonnegativity of quadratic forms defined on solutions of a first order eqaution*, Some Applications of Functional Analysis to Problems in Mathematical Physics (S. K. Godunov, editor), Inst. Mat. Sibirsk. Otdel. Akad. Nauk SSSR, Novosibirsk, 1988, pp. 97–100. (Russian)
34. ______, *An analog of the Weyl decompositin for first order operators*, Problemy Mat. Anal. **11** (1990), 46–50; English transl. in J. Soviet Math. **64** (1993), no. 6.
35. Yu. G. Reshetnyak, *Stability theorems in geometry and analysis*, "Nauka", Novosibirsk, 1982; English transl., Kluwer, Dordrecht, 1994.
36. V. A. Solonnikov and N. N. Ural′tseva, *Sobolev spaces*, Selected Chapters in Analysis and Higher Algebra, Leningrad Univ., Leningrad, 1982, pp. 129–196. (Russian)
37. H. Triebel, *Interpolation theory, function spaces, differential operators*, VEB Deutscher Verlag. Wiss., Berlin, 1977, and North-Holland, Amsterdam, 1978.

Selected Titles in This Series

(Continued from the front of this publication)

(See the AMS catalog for earlier titles)